The Physics of Basketball

The Physics of BASKETBALL

John J. Fontanella

The Johns Hopkins University Press
Baltimore

Printed in the United States of America on acid-free paper
9 8 7 6 5 4 3 2 1

The Johns Hopkins University Press
2715 North Charles Street
Baltimore, Maryland 21218-4363
www.press.jhu.edu

Library of Congress Cataloging-in-Publication Data
Fontanella, John (John J.), 1945–
The physics of basketball / John J. Fontanella.
p. cm.
Includes bibliographical references and index.
ISBN 0-8018-8513-2 (hardcover : acid-free paper)
1. Physics. 2. Basketball. 3. Force and energy. 4. Human mechanics. I. Title.
QC26.F66 2007
796.32301′53—dc22 2006010575

A catalog record for this book is available from the British Library.

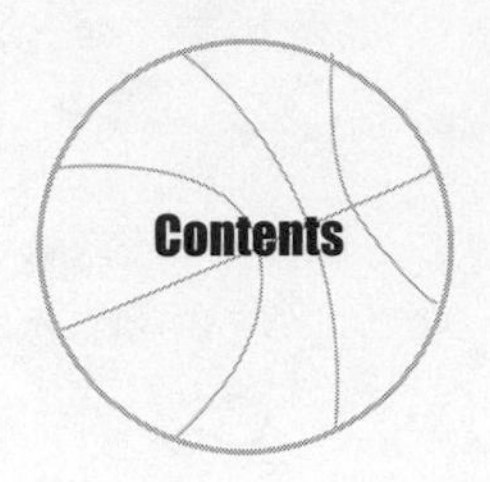

Contents

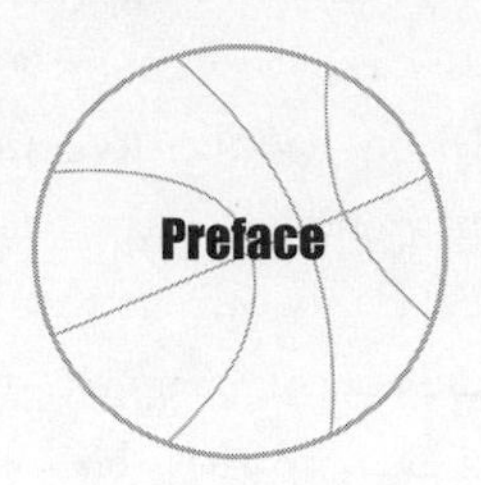

Preface

My brother will confirm that I began talking about writing a book on basketball and physics 40 years ago. I thought about it for 39.5 years, then started. I'm still not sure why I started even then. It might have been because of what I heard at some of the sessions on "Physics of Sports and the Human Body" during the 2004 Summer Meeting of the American Association of Physics Teachers. Maybe it was because of the U.S. basketball debacle at the 2004 Summer Olympics. Some of what I saw and heard made me think that I might have something useful to say about the game.

I was lucky enough to have had three outstanding coaches in my basketball career. My coach from grade school through sophomore year at Wampum High School, L. Butler Hennon, won 12 straight Section 20 Championships and 3 Class B State Championships and is a legend in western Pennsylvania. I was the starting point guard on the twelfth section championship team. Wampum High School was broken up after the 1961 season and I was sent to Mohawk High School, where John Samsa was coach. He was an excellent game coach. My coach at Westminster College, C. G. "Buzz" Ridl, was 1962 Coach of the Year of the National Association of Intercollegiate Athletics. He was 1974 Coach of the Year of the National Collegiate Athletic Association in the East while at the University of Pittsburgh. Presumably, Pitt grew tired of losing to teams coached by Buzz Ridl, so they hired him after the 1968 season.

As I began to think about some of their teachings from a physics point of view, it became clear why what they taught was correct. It also became clear that some of those lessons had been forgotten. There is also the matter of timing. I was fortunate enough to have been granted a sabbatical by the United States Naval Academy. Still another possible reason I started the book is that I've recently become enthusiastic about making physics relevant. Whatever the reason or reasons, this book represents my attempt to say a few words about some things that I love at a level exceeded only by my love for my family.

My hometown, Wampum, is a small town in western Pennsylvania. Even though western Pennsylvania is well known these days for football and football quarterbacks in particular, basketball was at least as important in the fifties and sixties. I grew up surrounded by athletes who became famous in other sports and who honed their skills via basketball. For example, I remember playing basketball against Joe Namath in 1961. It was a scrimmage between Wampum and his high school, Beaver Falls, which is about 10 km (6 miles) from my home. The list of my childhood heroes, all of whom grew up within about 30 km (20 miles) of my home includes Mike Ditka, Tito Francona (Terry's dad), Chuck Tanner, and Mike Lucci. About as close as we came to basketball fame is that Pistol Pete Maravich was born in the town where Mike Ditka grew up. Pete and his dad were legends in western Pennsylvania, but I didn't know until researching this book that he moved away when he was very young.

Nice discussions of some of the local sports figures of that era can be found in the books by Bob Vosburg, *Scooter's Days... and Other Days* (New Wilmington, PA, 1997) and *This Man's Castle* (New Wilmington, PA, 2005). Wampum is famous as the home of the Allen brothers, Coy, Caesar, Hank, Dick, and Ron. Hank, Dick, and Ron spent a total of 40 years playing professional baseball. The best known is Dick Allen who was Rookie of the Year in the National League in 1964 and Most Valuable Player of the American League in 1972. What he endured during his baseball career, as described in his autobiography, *Crash: The Life and Times of Dick Allen,* by Dick Allen and Tim Whitaker (New York, 1989), makes me doubly proud to have known him. My most vivid memory of Dick Allen is what he did during a basketball game at New Wilmington High School in 1960. I was a freshman sitting in the stands having just finished playing in the Junior Varsity game. I remember someone throwing a ball from about half court that was headed for a point about halfway up the backboard on the right. He jumped up to meet the ball then guided it down through the hoop. It was then that I knew that I would be playing the game just for fun. I was closest to his brother Ron who was the star on our twelfth section championship team.

I started college in 1963 in New Wilmington, another small town in the same county as Wampum. In 1962 Westminster had been named the top small college basketball team in the nation by both the Associated Press and the United Press International. One of the stars of that team, Ron Galbreath, was from Wampum. Ron became a college coach and was 10 wins away from 600 as of the writing of this book. Also, my neighbor, Mike Swanik, had been an outstanding contributor to another Westminster team, and thus going to school there seemed the obvious thing to do. Playing at a college close to home gave my parents and my uncle, Joseph Mikita, the opportunity to attend all of my home games and most of the away games. It made the successes more satisfying and the failures less painful.

Things were different then. The distinction between the basketball teams from small and large colleges wasn't quite so pronounced as it is today. Very early on there was no difference. For example, according to *Madison Square Garden* by Zander Hollander (New York, 1973) and *The Story of Basketball* by Lamont Buchanan (New York, 1948), on December 29, 1934, Westminster defeated St. John's 37–33 during the first basketball doubleheader played at Madison Square Garden. That tradition continued through when I played. I took great pleasure in beating Pitt during both my junior and senior years. The win over Pitt during my junior year alerted Syracuse that we were a decent team so they were ready for us when we visited there to end the season. We were beaten badly, but the game almost gave me a claim to fame since I outscored Dave Bing 24 to

22. I say "almost" because Bing left at half-time to play in an All-Star game. Jim Boeheim, the current coach at Syracuse, had 13 points in the game for Syracuse. A book has recently appeared describing those times, *The Glory Days* by Dick Minteer (Westminster College, 2004). For Westminster, that era ended in 1969 which was the last year that they defeated a Division I school.

I must admit that the transition from basketball player to physicist was difficult. It was my own fault. I had spent huge amounts of time playing basketball. I must have been trying to evaluate one of the statements that I recently noticed in Bob Vosburg's first book: "Hard work will beat talent if talent doesn't work hard." Fortunately, the people at Case were patient and understanding. Thanks to them and further testing of the quote from Vosburg's book, I've been able to spend an enjoyable career teaching and doing research at the Naval Academy.

By now I've spent a lot more time teaching and doing physics than playing basketball. Consequently, the book draws on extensive time spent in both the scientific and athletic communities. This duality makes it difficult to categorize the book. One possibility is that the book is basketball from a physics point of view. Another is that it is physics from a basketball point of view. I like to think that it's both. On the basketball side, I wanted to say some things that might give a basketball player an edge. Even a rudimentary understanding of what is possible and what is not can significantly streamline and accelerate learning the game. The need for this has taken on increased importance in today's world of video games and movies where reality is often not on display. It is also possible that an advanced understanding of the underlying physics can help distinguish between subtle differences in technique. The other part of what I wanted to say has to do with the physics itself. I wanted to communicate some of the physics in action I now see in the game of basketball. I hope that there is some information in the book that physics teachers can use in the classroom.

Though it is difficult to say what the book is, it is clear what the book is not. It isn't a novel. While sections of the book should be pleasant reading, a large fraction of the book is technical. It describes the hows and whys of basketball and uses physics to distinguish good technique from bad. Further, there is no attempt to show how to teach the techniques that are identified in the book as best. I tried some coaching of my son's and daughter's basketball and baseball or softball teams. It quickly became obvious that I should not quit my day job to become a coach. Because of those experiences, there is no advice in the book on how to coach. Fortunately, there are some excellent books on coaching. One of the best is the book by Morgan Wootten, *Coaching Basketball Successfully* (Champaign, IL, 2003). Because of that book, and because I have finally gotten around to thinking through the game, it might be fun to try coaching again.

Finally, because of the work that went into this book, I watch basketball games differently now. Who does what and the flow of the game are still the focus. However, I find myself thinking more about how players do what they do and evaluating how well they do it. The flight, spin, and bounce of the ball and the motion of the players and the interactions that they cause or experience have all become more important. I enjoy watching the game more now though I'd still rather be a doer than a viewer. My final hope, then, is that some of the insights contained in the book will deepen the appreciation of the game for the serious fan.

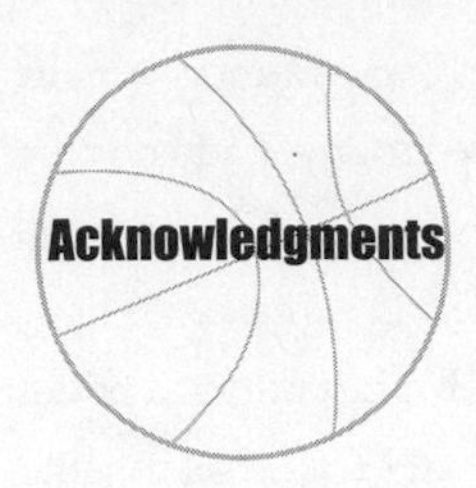

Acknowledgments

Writing this book has been a family effort. My wife, Dr. Mary Wintersgill, is a physicist and is currently chair of the physics department at the United States Naval Academy. She has been a great help with all aspects of the book. My son has just entered seminary. He must have contributed since it's clear that the book has made it into print only because of divine intervention. My daughter is an English major and physics minor at college and made extensive suggestions for improving all of the chapters. In fact, all three family members suffered through innumerable versions of the manuscript.

Special thanks go to Mrs. Betty Ridl for providing a great deal of useful information. In addition to her many activities at Westminster College and the University of Pittsburgh, Mrs. Ridl serves on the Advisory Board for Coaches vs. Cancer. Joe Onderko, the sports information director at Westminster, excavated some ancient statistics for me. I also thank my many teammates. What I didn't learn from my wonderful coaches, I learned from them. My contact for basketball at USNA has been Dave Smalley. During his career he successfully coached the men and started the women's program. Dave has a great deal in common with my college coach. They are examples of masters and gentlemen of the game. Many others have recognized that. The gym at Westminster is known as Buzz Ridl Gymnasium and the varsity basketball court at the Naval Academy is named Dave Smalley Court.

I am indebted to my undergraduate school, Westminster College (New Wilmington, Pa.), for providing an atmosphere where an athlete can also be a student. I am also indebted to my graduate school, Case, for providing an opportunity for someone who was initially more athlete than student. My thesis advisor at Case, Donald E. Schuele, has had a lifelong interest in sports. I met Don on the basketball court during a noontime pickup game and we played basketball, softball, and volleyball together in various leagues. Don served on the Science and Engineering Technology Committee of the U.S. Olympic Committee for 10 years.

I have been fortunate to have had a succession of outstanding research collaborators: Dr. Carl G. Andeen of Andeen-Hagerling, Inc., Dr. Alan V. Chadwick of the University of Kent at Canterbury (U.K.), Dr. Steven G. Greenbaum of Hunter College of CUNY, John T. Bendler, Donald J. Treacy, and Charles A. Edmondson of the USNA Physics Department, and Michael F. Shlesinger of the U.S. Office of Naval Research. I have learned a lot about science from them. Thanks also go to Kevin Sinnett of USNA for providing me with a list of his favorite moments in basketball history. For commenting on various parts of the manuscript I am grateful to Larry Ondako, the men's basketball coach at Westminster; Ron Galbreath, the women's coach at Geneva College; Dave Beam of Beam & Associates; and Don Schuele, John Bendler, Mike Shlesinger, and Frank Gomba. I also thank Dave Rector of USNA for arranging for me to use the LoggerPro® software, Vernier force plate®, and a LabPro® data acquisition device. Finally, I am indebted to Trevor Lipscombe, editor-in-chief of the Johns Hopkins University Press, for his interest in the book. Trevor made several very helpful suggestions during the preparation of the manuscript.

The Physics of Basketball

The Final Four

Physics: It's better than you think. It has to be. That sentiment was borrowed from the Baltimore Opera Company. It's what they say about opera in their TV and radio commercials. I suppose that writing about opera and physics is not a good way to begin a book. I would like more than five people to read it.[1] However, opera definitely belongs in a book about basketball since one of the great sports quotes of all time, "The opera ain't over until the fat lady sings," became famous because of a basketball game.[2] Besides, physics and opera have a lot in common. They are both about truth and beauty. That's what this book is all about—truth and beauty in the game of basketball. Let's get started.

All aspects of the game of basketball are controlled by forces. Without forces, a basketball or a basketball player would always move in a straight line with a constant speed. A special constant speed is zero. In that case, the basketball or basketball player would always be at rest. These are consequences of an important law of physics, Newton's First Law (N1L). Needless to say, without forces there would be no game, so we'll start with a general discussion of forces and what they do. If you already know a lot about forces, at least those usually presented in a typical general physics course, you might consider skipping to the last section of this chapter, "A Little Different Spin on Things."

There's a lot of insight into the nature of forces that can be gained from these common sentences or phrases: "may the force be with you," "it was a force-out at second base," "police force," and "she was a force in the senate." Each of the examples hints at the scientific definition of force, since each indicates that a force is involved when one thing "influences" another. The first thing that we must decide is what a force *is*. It turns out that we can describe the game of basketball by defining a force to be a push or a pull. Although basketball is not usually thought of as a game of contact, pushes and pulls (forces) by one player on another are part of the game. The pick and roll is not something that you have for breakfast. I've seen some violent collisions happen during a pick. The idea is to get in the way of someone trying to guard a teammate. Picks are most effective when the person on defense is not aware that it's going to happen. The rules make the violence worse since the person setting the pick must be firmly in place.

Some players go beyond the acceptable pushes and pulls. The most notorious "bad boy" was Bill Laimbeer who is now coach of the 2003 Women's National Basketball Association champion Detroit Shock. Laimbeer played for the Detroit Pistons for most of his 15-year career that began in 1980. Laimbeer was well known for his use of "unauthorized force." Before Laimbeer it was Norm Van Lier who played with the Chicago Bulls for several years during the seventies. I can attest to Van Lier's "aggressiveness." We played against Van Lier because his college, Saint Francis University (Pa.), was in Westminster's conference. I remember being to the right of the foul line during a game in February 1967 when I looked up and saw Van Lier coming at me with a menacing look and his fist up. Now, because of basketball, I've been pushed, elbowed, and defeated, but I've never been intimidated. I smiled, bent over a bit, and drove my bony right shoulder into his gut. I straightened up a bit and Van Lier became a quick study in projectile motion. He didn't bother me again.

Getting back to the physics, one of the forces that physicists talk about is the force of gravity. The phrases "pull of gravity" or "gravitational pull" are familiar and are consistent with our definition of a force. Other examples are "push a car" (I drive old British sports cars so I do that a lot), "pull a wagon," or the "pushing" violation (foul) in basketball.

The next thing that we need to decide is what a force *does*. We already know that without forces a basketball or a basketball player would always move in a straight line with a constant speed. A physicist would say that if something moves in a straight line with a constant speed it has a constant velocity. It follows that what a force *does* is change the velocity of a basketball or basketball player. This is a case of cause and effect. It is reasonable to say that a force is a cause and a change in velocity is an effect. This makes clear the role of forces in the game of basketball. Both the ball and (a good) player are perpetually changing direction and/or speed and the only way that that can happen is via forces.

Physicists have a name for what happens when the velocity changes and that is *acceleration*. Acceleration occurs when there is a *change* in either the speed or direction of travel of a moving object. Both could change simultaneously. Just as velocity tells us how fast the position changes, the acceleration tells us how fast the velocity is changing. (A physicist would define velocity as the rate of change of position and would define the acceleration as the rate of change of velocity.) A familiar example is an airplane accelerating down the runway during takeoff. It is accelerating because it is speeding up. (Note: If we are sitting in an airplane seat during takeoff, what we feel is the force of the seat pushing us in the forward direction. The acceleration is the rate change of velocity. Once the distinction between force and acceleration becomes clear, many of the mysteries of classical physics, at least, disappear.) The best way to observe acceleration is to watch a video of Michael Jordan, arguably the best basketball player ever. Jordan played for the Chicago Bulls from 1984 to 1998 then finished his career with the Washington Wizards during the 2002 and 2003 seasons. What MJ did best was accelerate. In my opinion, his ability to change direction and change speed was unsurpassed.

To describe what made MJ play and to describe the physics of basketball in general, we need to define one more quantity, mass. Mass *is* the amount of matter that an object possesses. For example, 2.16 m (7′1″) tall

Shaquille O'Neal, currently playing for the Miami Heat, has more mass, 148 kilograms (10.2 slugs) than 1.6 m (5′3″) tall Tyrone "Muggsy" Bogues, 63.2 kilograms (4.4 slugs). Muggsy played college basketball at Wake Forest University and is the shortest player ever to play in the National Basketball Association. He had a 14-year NBA career including nine-plus years as a Charlotte Hornet and is currently the coach of the WNBA Charlotte Sting. Since they are made up of the same kind of stuff, Shaq has more mass than Muggsy because there is more of Shaq.

What mass *does* is oppose acceleration. A force on something with a large mass will result in a smaller acceleration than if the same force acts on a smaller mass. This makes sense since a force on Shaq will have much less effect (cause much less acceleration) than the same force on Muggsy.

The relationship between force, mass, and acceleration is the essence of another important law of physics, Newton's Second Law (N2L). The usual statement of N2L is that the total (net) force on an object equals the product of the mass times the acceleration of the object. More than one force can act on an object (e.g., Shaq, Muggsy, or MJ) at the same time. Those forces can be added so long as we are careful to include both the strength (magnitude) and direction of the forces. The total (net) force is just the sum of the forces.

It's time for our first equation. Suppose that we represent the forces on an object by $\mathbf{F}_1$, $\mathbf{F}_2$, and so on. The boldface indicates that a force has both strength and direction. A physicist or mathematician would call $\mathbf{F}_{gravity}$ a vector. If m is the mass of the object and $\mathbf{a}$ is the object's acceleration, Newton's Second Law is usually written as

$$\mathbf{F}_1 + \mathbf{F}_2 + \ldots = m\mathbf{a}. \qquad (1.1)$$

This implies in symbols what we said in words. For a given set of forces, $\mathbf{F}_1 + \mathbf{F}_2 + \ldots$, if the mass is small, $\mathbf{a}$ is large and vice versa. It is easier to accelerate Muggsy than it is to accelerate Shaq.

Equation (1.1) is not the way N2L was originally stated by Newton. It is a special case that only applies to situations where the mass is not changing.

For example, it is not useful for rockets where the mass decreases as the fuel is burned. However, equation (1.1) works just fine for our purposes. It accurately describes the relationship between forces and accelerations for basketballs and players relative to a basketball court on the surface of the Earth.

There is often a lot of nonsense associated with discussions of N2L. Force is often confused with acceleration. Part of the reason is that it is sometimes said that "Force *is* mass times acceleration." That is totally bogus. N2L says that force is *mathematically equal to* the mass times the acceleration. However, force is something different from acceleration and mass is something different from both force and acceleration.

The Gravity of the Matter

Let's consider a falling basketball in some detail. The most important force on a basketball falling at a speed typical of those in a game is the force of gravity. Gravity is the force by which masses attract one another. Gravity is a fundamental force. It cannot be broken down into other forces. None of the other forces described in this book are fundamental forces. The gravitational force is sometimes known as the weight. On the surface of the Earth, the gravitational force on Shaq is 1,450 newtons (325 pounds) and the gravitational force on Muggsy is 619 newtons (141 pounds). If we ever get to play basketball on the Moon, our weight there will be a lot less, for the Moon's gravitational pull is only about a sixth of the Earth's.

Unfortunately, confusion is often produced by the phrases "gravitational acceleration" or "acceleration of gravity." It is important to realize that an object has "gravitational acceleration" if the velocity of the object is changing only because of the force of gravity. That happens if gravity is the only force that acts on the object. However, the force of gravity acts on an object near the Earth whether or not it is accelerating. Consider a basketball sitting on the floor. Gravity, the force, still acts on the ball, yet there is no acceleration because the velocity is not changing. The reason that the acceleration is zero is that the total (net) force on the ball is zero. The total (net) force on the ball is zero because, in addition to the downward force of gravity, there is an upward force on the ball when it is sitting on the floor, the force of the floor on the ball. The force of the floor on the ball is equal and opposite to the downward force of gravity.

Beware—that the downward force of gravity is equal and opposite to the upward force of the floor on the ball is *not* a consequence of Newton's Third Law (N3L). The usual (and misleading or incomplete) statement of N3L is for every action there's an equal and opposite reaction. What is usually missing is that the action and reaction forces must act on *different* objects. The reason that the force of gravity and the force of the floor are *not* action and reaction forces is that they both act on the basketball.[3] If we think of both of those forces as action forces, there are two other (reaction) forces that act on something else. They are easy to find. Consider the gravitational pull of the Earth (down) on the ball. The corresponding reaction force is the gravitational pull (up) of the ball on the Earth. That sounds strange since it is unsettling to think that the ball pulls on the Earth. What is also strange but true and a consequence of N3L is that if Shaq runs into Muggsy, the force of Muggsy on Shaq is equal (and opposite) to the force that Shaq exerts on Muggsy. Getting back to the basketball, the reaction force to the force of the floor on the ball (up) is the force of the ball on the floor (down).

Sorry about the detour. Let's get back to the effect of gravity on the basketball. We will represent the force of gravity on the ball by an arrow and label it $\mathbf{F}_{gravity}$. The force of gravity on a basketball near the surface of the Earth is shown in figure 1.1.

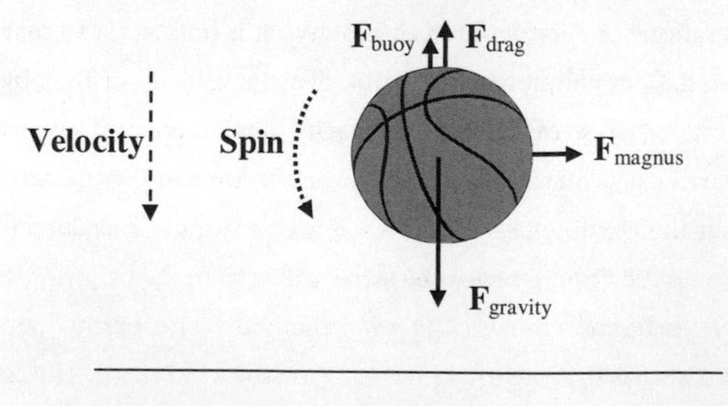

Figure 1.1. The forces on a falling, spinning basketball in air near the surface of the Earth

The direction of the force of gravity on the ball is downward since the Earth pulls the basketball downward. When drawing vectors, I like to put the tail of the arrow at the point where the force is applied. That would make the vector appear to originate where the force is applied or appears to be applied. The tail of the arrow $\mathbf{F}_{\text{gravity}}$ is drawn at the center of the basketball in figure 1.1 since that's close to where the total (net) gravitational force on the ball appears to act, the so-called center of gravity. As we'll see, sometimes it's not convenient to draw the tail of the arrow at the origin of the force.

The strength of the force of gravity on the ball is also known as the weight of the ball, so we know that the strength of the force of gravity on a basketball near the surface of the Earth is about 6 newtons (1.3 pounds). The strength of the force of gravity (weight) on an object can be predicted using Newton's Law of Universal Gravitation. That law gives the following equation for the strength of the force of gravity on a basketball near the surface of the Earth.

$$F_{\text{gravity}} = \frac{G m_{\text{basketball}} m_{\text{Earth}}}{R^2_{\text{Earth}}} \tag{1.2}$$

G is a universal constant (6.67×10^{-11} newton meter^2 per kilogram^2), $m_{\text{basketball}}$ is the mass of the basketball, m_{Earth} is the mass of the Earth and R_{Earth} is the radius of the Earth. We are assuming that the basketball is a distance R_{Earth} from the center of the Earth. Equation (1.2) can always be used to calculate the force of gravity. Plugging values for G, and the mass and radius of the Earth into equation (1.2), we get $F_{\text{gravity}} = m_{\text{basketball}}$ times 9.8 meters per second^2 (= $m_{\text{basketball}}$ times 32 feet per second^2). For most of the remainder of the book, I'll use m for meters, ft for feet, in for inches, s for seconds, N for newtons, lbs for pounds, and kg for kilograms. Notations such as meters per second^2 will be written as m/s^2 though miles per hour will be written as mph.

Confusion arises because 9.8 m/s^2 has units of acceleration and is often referred to by the symbol g ($g = Gm_{\text{Earth}}/R^2_{\text{Earth}}$) and the names "gravitational acceleration" or "acceleration of gravity." However, the way that we have used it so far, 9.8 m/s^2 is not acceleration. It is just a number that we can multiply mass by to get the force of gravity (weight). For example, equation (1.2) is usually rewritten as

$$F_{\text{gravity}} = m_{\text{basketball}}\, g \tag{1.3}$$

As has been mentioned, there is one special case where 9.8 m/s^2 is an acceleration. That special case is when gravity is the only force on an object near the surface of the Earth. Since the force of gravity is proportional to the mass of the object, the acceleration does not depend on the mass of what is falling. This implies that the acceleration of either a men's or a women's basketball falling under the influence of only gravity would be 9.8 m/s^2. That would be the case if a basketball were falling in a vacuum and would happen if the Earth had no atmosphere or air.

If a falling basketball had an acceleration of 9.8 m/s^2 it would be speeding up 9.8 m/s every second. For example, a basketball starting from rest would have a speed of 9.8 m/s after one second, 19.6 m/s after two seconds, and so on. Where the basketball is located during the fall is slightly more complicated. Since the ball speeds up as it falls, the distance that it travels in equal time intervals increases as time passes. For example, during the first second, the ball falls 4.9 m (16 ft), during the second second it falls 14.7 m (48 ft), and so on.

By the way, if you are in the mood sometime to really annoy someone, ask a physicist what causes gravity and how gravity can pull on an object when the object is in the air. My suggestion is that you do not listen too closely to the answer if it's anything other than "We're not really sure but here are a few ideas . . ."

This has been a long discussion of gravity, partly as a warmup and partly because gravity is the dominant force on a ball (or player) in flight at the usual speeds associated with a typical basketball game. Since gravity is the main factor influencing a shot, passed basketball, or an airborne player, it can be said that the ability of players to deal with gravity is the main factor that determines their level of success. We focus mostly on shooting, because it's fun. Remember, though, that gravity certainly affects passing. The Saint John's fans who went to Madison Square Garden in New York and saw Chris Mullin's skills know what a fine art passing is, even if Chris' teammate Bill Wennington often failed to anticipate the arrival of the ball. Bill ended up with several rings from his professional career as a benchwarmer for the Bulls while Mullin remained ringless with the Golden State Warriors.

In addition to gravity, there are three other forces on a basketball falling through the air that we need to consider. If the game were played where there is no atmosphere or air, the other three forces wouldn't exist and a ball or person in flight would indeed only be affected by gravity.

It's a Buoy

The first force caused by air that we will consider is the buoyant force, $\mathbf{F}_{buoy}$. Because a basketball is relatively large and has a small weight, the buoyant force is nonnegligible. As is also true for the gravitational force, the buoyant force acts on a basketball whether the ball is moving or not. The buoyant force is the force of Archimedes's fame and is what makes boats float or helium balloons rise. In the case of a floating boat at rest, the buoyant force is the total (net) upward force of water molecules on the boat. In the case of a helium balloon or basketball at rest, the buoyant force is the total upward force caused by collisions of the air molecules with the outside surface.

The buoyant force is shown in figure 1.1. I would have preferred to have drawn it at the center of the ball since that is close to the center of buoyancy where the buoyant force appears to act. Since that is not convenient, $\mathbf{F}_{buoy}$ is drawn at the top of the ball. We do gain something by drawing it with the tail at the surface because that reminds us that it is caused by collisions of the air molecules with the surface.

Had Archimedes played basketball, he would have said that the buoyant force is equal to the weight of the air that the basketball displaces, that is, the air that isn't there because the basketball is. He would have said that because he realized that if the basketball weren't there, that portion of the air would be in equilibrium on average. Since there is a downward force of gravity on that portion of the air, there must be an equal and opposite force upward on that portion due to the rest of the air. The upward force due to the rest of the air is the buoyant force. The same upward buoyant force is exerted on the basketball by the rest of air, because the rest of the air doesn't know or doesn't care what occupies the volume.

We can calculate the upward buoyant force because we can calculate the weight of the displaced air. We know the volume of the displaced air since it is equal to the volume of a basketball, $V_{basketball}$. We also know the density of air, ρ_{air}. Consequently, $\rho_{air} V_{basketball}$ is the mass of the displaced air. Equation (1.3) tells us to multiply this by g to get the weight of the displaced air. According to Archimedes, this is equal to the strength of the buoyant force so we can write the following equation:

$$F_{buoy} = \rho_{air} V_{basketball}\, g. \tag{1.4}$$

Calculations based on equation (1.4) show that for both men's and women's basketballs, the buoyant force is about 1.5% of the weight. Because the buoyant force is upward, it can be subtracted from the force of gravity that is downward. Consequently, the total force on a basketball due to gravity and buoyant force is downward and equal to about 98.5% of the weight. This has the effect of changing the "gravitation acceleration" of a basketball from 9.8 m/s^2 to about 9.66 m/s^2. A basketball under the influence of only gravity and the buoyant force would fall with an acceleration of about 9.66 m/s^2. Since the buoyant force on a basketball acts to just slightly reduce the effect (force) of gravity, the shape of the trajectory of a thrown basketball under the influence of only gravity and the buoyant force would be the same as if it is only affected by gravity. The other two main forces of the air on the ball do change the shape of the trajectory.

By the way, the reason that a helium balloon rises is that its weight is less than the buoyant force. Since the upward buoyant force is greater than the downward force of gravity, a helium balloon affected only by gravity and the buoyant force would accelerate upward.

What a Drag

The second force on a basketball caused by air is the drag force, which is sometimes known as air resistance or air friction. The drag force acts on a basketball only when it is moving. What happens is that, to move through the air, a basketball must "force its way" forward; it must push the air molecules out of the way. By N3L, the air molecules must exert a reaction force backward on the basketball. The reaction force of the air is the drag force. Because the reaction force is backward, the drag force is in a direction opposite to the velocity of the basketball. In figure 1.1, the drag force, $\mathbf{F}_{drag}$, is represented by an arrow pointed upward since the ball is traveling downward. If the basketball had been moving upward, the drag force would be downward, in the same direction as gravity.

For the speeds typically encountered in a basketball game, we can write an equation that describes the drag force on a basketball. We can do this because the force on a sphere traveling through air at intermediate speeds is well known. This is an unusual situation in which a sphere is actually a reasonable approximation of what we are interested in—a basketball. This never stops a physicist, however, since there is some truth to the old (and not very good) joke about the physicist who was asked to describe a chicken. The response of the physicist was "Assume a spherical chicken . . ."

The first important characteristic of the drag force is that the faster a basketball moves, the larger the strength of the drag force. For the speeds typical of a basketball game the air drag force on a basketball varies as the square of the speed, v. The drag force is also proportional to a special area, A. This is the area that the basketball "sweeps out" as it moves through the air. Finally, the drag force is proportional to the density of the air. The more dense the air, the more difficult it is for the basketball to make its way through. This gives us the following equation for the strength of the drag force:

$$F_{drag} = C_{drag}\,\rho_{air} A v^2 \tag{1.5}$$

C_{drag} is a constant. The details of the application of equation (1.5) are given in appendix I.

We are in good company in the use of equation (1.5) since it was quoted by Newton in 1687.[4] A plot of the drag force versus speed on a men's basketball is shown in figure 1.2 along with the constant force of gravity. Because of the smaller size of a women's basketball, the drag force is about 7% lower than the drag force on a men's basketball. For speeds at which the game is normally played (less than about 10 m/s or 22 mph), the drag force on a basketball is less than about 15% of the force of gravity. This is small, but cannot be ignored as we will see. If the basketball is traveling fast, the drag force can be large.

The graph in figure 1.2 shows the drag force for speeds higher than are usually achieved in a basketball game to point out the special speed (about 21 m/s or 47 mph) known as the terminal speed. When a basketball is traveling downward at the terminal speed the strength of the (downward) force of gravity equals the (upward) drag force for a falling object. We are ignoring the buoyant force. The value of the terminal speed would be slightly less if the buoyant force were included. The terminal speed is the fastest that a men's basketball dropped from a bungee tower, for example, would travel.

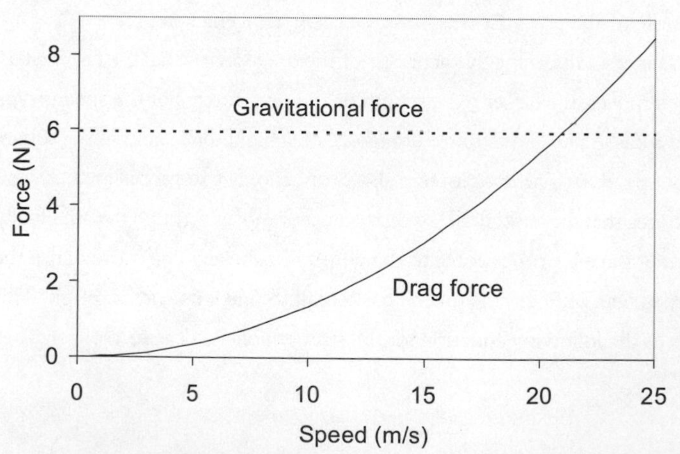

Figure 1.2. Plot of the theoretical drag force and the gravitational force versus speed for a falling men's basketball near the surface of the Earth

What happens is that as the basketball speeds up as it falls, the drag force increases. The strength of the drag force gets closer and closer to that of gravity and thus the net force gets closer and closer to zero. This makes the acceleration of the ball get closer and closer to zero. Consequently, the ball speeds up at a slower and slower rate and the speed gradually approaches 21 m/s. A dropped men's basketball never actually reaches 21 m/s but it gets close enough for all practical purposes.

This does not imply that a basketball cannot travel faster than the terminal speed. For example, there might be a third force on the basketball, such as someone throwing the ball downward from the bungee tower or a "basketball gun" so that the speed of the basketball when it comes off of the bungee tower is faster than 21 m/s downward. This is not too far-fetched because such velocities can be involved when a basketball is thrown the length of the court. What happens if a basketball leaves the bungee tower with a speed greater than the terminal speed is that the ball slows down to a speed of 21 m/s.

In the next section of this chapter we consider the effect of spin on a ball. There should also be a drag force tending to slow down the spin of the ball. I carried out a large number of experiments to determine the spin of a basketball during a typical shot. In all cases, the spin of the basketball remained constant to within the uncertainty in the experiments. Consequently, the drag force tending to decrease the spin of a basketball appears to be very small.

A Little Different Spin on Things

The final important force on a basketball traveling through the air is the Magnus force.[5] For this force to act, the basketball must be both moving and spinning. For example, if the ball has a velocity downward and is spinning counterclockwise the Magnus force is to the right as shown by the force labeled $\mathbf{F}_{magnus}$ in figure 1.1. The Magnus force is the force that makes a baseball curve. In the book, *The Physics of Baseball,*[6] Adair points out that the force due to air drag is different on opposite sides of a spinning ball moving through the air. The reason is that the total speeds of the surfaces of the ball are different. For example, in figure 1.1, the total speed of the left surface of the basketball is greater than the total speed of the right surface of the ball. That is because the left surface is moving downward relative to the center of the ball which is moving downward. Consequently, the total speed of the left surface is the sum of the two speeds. On the right of the ball, the surface is moving upward relative to the center of the ball so that the total speed is the difference. Since the drag force is larger when the speed is larger, it follows that the drag force on the left side of the basketball is greater than the drag force on the right. This gives rise to an extra force to the right, the Magnus force.

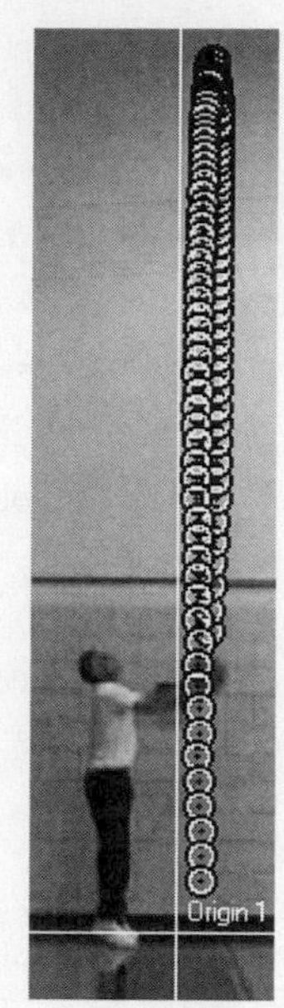

Figure 1.3. Path of a ball thrown upward with counterclockwise spin

Like the other forces caused by the air, the Magnus force on a basketball is usually small. In fact, I was skeptical that it is important at all until I made the video represented in figure 1.3. The video was shot using a Sony DCR-HC20 Digital Video Camera Recorder and was transferred to a computer via VideoPoint® Capture 2.0 software. VideoPoint® 2.5 was then used to determine (digitize) the position of the basketball every 1/60 second. The open (white) circles in figure 1.3 represent the digitized positions of the ball. The ball is spinning counterclockwise at about 4 revolutions per second. On the way up (the partially hidden circles) the ball is initially also moving to the right slowly and then begins moving to the left. This indicates that, on the way up, the basketball has some acceleration to the left. This requires a force to the left. The force to the left is provided by the Magnus force. On the way up, the basketball is also slowing down (accelerating downward) because of the force of gravity. On the way down (the full circles), the ball starts out moving to the left and about halfway down begins moving to the right. The curve on the downward flight is easily seen by comparing the path of the ball with the vertical line. This indicates that, on the way down, the basketball has some acceleration to the right. The required force to the right is provided by the Magnus force. On the way down, the basketball is also speeding up (accelerating downward) because of the force of gravity. In other words, a ball with spin thrown upward travels in a tiny loop. The loop is small since the deflection from about halfway down to the bottom is only about 5 cm (2 in). This shows that the Magnus force is small but observable in this case. However, the Magnus force does affect the game. In fact, as we will see, there can be circumstances in which the Magnus force is large.

The next thing that I was skeptical about is that an equation can be written for the Magnus force on a basketball. I was worried because, at the linear and rotational speeds of a pitched or batted baseball, there does not appear to be a valid, simple theory.[7] It has been suggested recently, however, that the Magnus force is reasonably well understood for a soccer ball and a basketball under normal conditions.[8] According to Ireson,[9] the Magnus force is greater the faster the ball is traveling. We'll refer to that speed as the linear speed of the basketball. The linear speed is the speed of the center of the basketball and is the same speed, v, that was used in equation (1.5). The Magnus force is also greater the faster the spin on the ball. We will refer to the amount of spin as the rotational speed, ω. The rotational speed is measured in revolutions per second or rps. We could have used revolutions per minute or rpm but the numbers would be 60 times as large. The Magnus force also depends on the volume of the ball. We will write the equation in terms of the diameter, D, of the basketball. Finally, the Magnus force depends on the density of the air. Consequently, the strength of the Magnus force can be approximated by[10]

$$F_{\text{Magnus}} = C_{\text{Magnus}} \rho_{\text{air}} D^3 \omega v \tag{1.6}$$

C_{Magnus} is another constant called the Magnus coefficient. Further mathematical details are given in appendix II.

I used equation (1.6) to predict the path shown in figure 1.3. The theory did reproduce the loop reasonably well. However, the loop is very small and the difference between the theory and the experiment was sufficiently large that I still had some doubts about the accuracy of equation (1.6). More speed was necessary to make the Magnus force and hence the deflections larger. I went to Halsey Field House at USNA and threw a men's basketball in about the same way. However, in this case I was standing on the edge of a wall so that the ball continued to fall an extra 3 m (10 ft) and thus the ball continued to speed up after it passed me. Also, I threw the ball with both topspin (I was facing away from the edge of the wall and threw the ball back over my head) and backspin (I threw the ball as though I were shooting an old-fashioned underhand foul shot). I made a video of the throws and digitized them as described earlier. The horizontal position of the basketball versus time was determined from the data. The results are shown by the open symbols (squares for topspin and circles for backspin) in the graph in figure 1.4. Also shown are solid lines representing theoretical calculations based on equation (1.6). The calculation techniques used for the theory are given in appendix III. The rotational speed for the backspin plot is 4.5 revolutions per second and the rotational speed for the topspin plot is 2.5 revolutions per second.

The topspin curve curves down and the backspin curve curves up. If only gravity and the buoyant force were acting, each set of data in figure 1.4 would be a straight line. The reason is that both gravity and the buoyant force act only in the vertical direction. There is nothing affecting the basketball's horizontal motion so the horizontal speed would be constant. That would give a straight line in a plot of the horizontal position versus time. There is a small

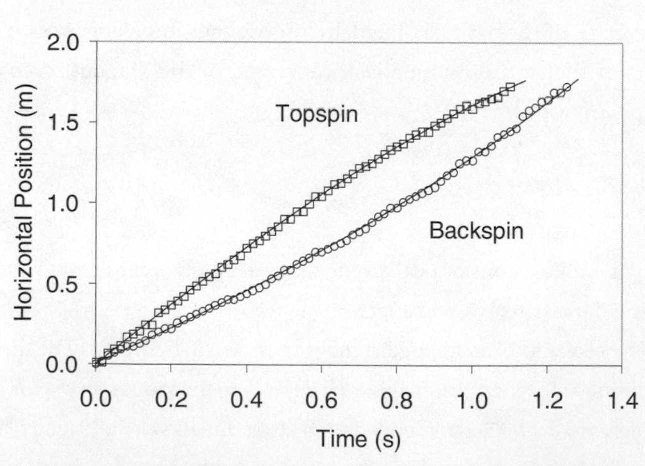

Figure 1.4. Plot of the horizontal position of a basketball thrown off a wall versus time. The data are shown by the open symbols and the lines are the theoretical curves.

amount of drag force but that should only slow the ball in the horizontal direction. That would make both curves curve down. It is clear that while the topspin plot is curved down, the backspin plot is curved up. In addition, there is good agreement between the theory (lines) and the experimental results (points). This finally convinced me that not only is the Magnus force important for a basketball but it can also be modeled via equation (1.6).

The model predicts that the Magnus force on a women's basketball is about 10% smaller than the Magnus force on a men's basketball. Equation (1.6) also predicts that for the same linear speed and rotational speed, the Magnus force on a basketball is about 38 times stronger than the Magnus force on a baseball. The large difference occurs, in part, because basketballs are much larger than baseballs and the Magnus force depends on the cube of the diameter. This prediction is suspicious because basketballs are not generally known to curve. The reason that the Magnus force is usually more important to the game of baseball is that both the linear and rotational speeds of a baseball are usually much greater than for a basketball. There are occasions, however, when the Magnus force is large enough for the curving of a basketball to be easily observed. I remember having to compensate for effects of the Magnus force when I threw the ball the length of the court. The Magnus force is large in that case because both the linear and rotational speeds of the ball are large, in particular, if one uses the sidearm technique that I was taught for throwing a basketball a long distance. When I was in grade school, Coach Hennon showed us that the easy way to launch a basketball a long distance is to throw it using a sidearm technique similar to that used in throwing a discus. The technique is demonstrated in figure 1.5.

When the ball is thrown in this manner by a right-handed player, it spins around an approximately vertical axis so that the ball curves to the right as viewed by the thrower. The spin and horizontal trajectory are similar to those for a screwball in baseball. The curving of the ball, caused by the Magnus force, can be seen in figure 1.5 by comparing the digitized positions of the basketball with the vertical white line. Just after the ball

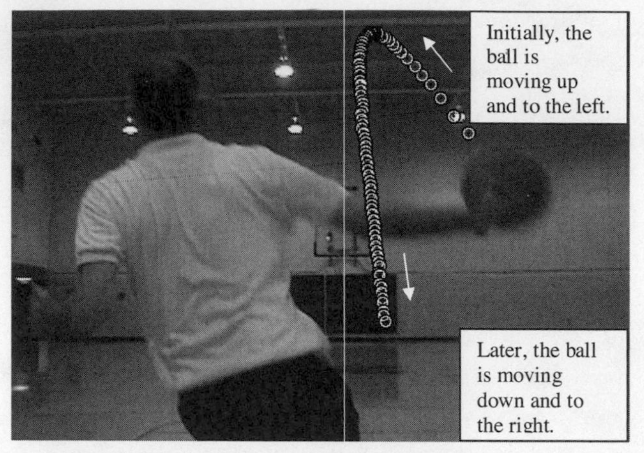

Figure 1.5. One technique for throwing a basketball a long distance as demonstrated by the author. The points show position of the ball spaced by 1/60 second.

leaves the hand it is traveling up and to the left. After reaching the peak of the trajectory, the ball is moving down and to the right. This is a technique that I haven't seen used much in the past few years, though it can be quite useful for those who don't have the strength to throw a "baseball" pass a long distance. I used this technique fairly frequently to throw the ball to teammates at the other end of the court. It has gotten my teams a lot of quick baskets over the years.

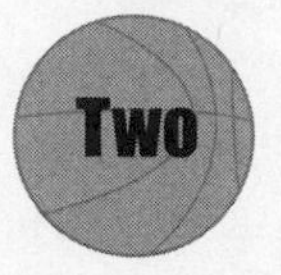

Projectile Notion

One of the greatest events in basketball history took place on March 2, 1962. On that night, Wilton Norman Chamberlain scored 100 points. Wilt was playing for the Philadelphia Warriors who defeated the New York Knickerbockers, 169–147, in a game played in Hershey, Pennsylvania. That night he broke his own record of 78 that he had set earlier in the season. Except for the few in attendance, the best that any of us can do is to read about the game[1] since it was not televized. When the Knicks recognized that Wilt was "on" that night (Today, we would say that he was "in the zone.") they began to stall and mob him with defense. Wilt said ". . . I maybe could have scored 140 if they had played straight-up basketball." The Knicks also tried to foul him since Wilt was known as a terrible free-throw shooter. That didn't work out for the Knicks because Wilt made an uncharacteristic 28 of 32 free throws. The Warriors countered by fouling the Knicks since that was the only way to get the ball back quickly. To their credit, the Warriors increasingly fed the ball to Wilt. That was the first thing that Wilt recognized. He said that "It would have been impossible to score this many if they hadn't kept feeding me." What Wilt says is true. Basketball really is a team sport. (It is sometimes said that basketball is a game, not a sport like track and field. We ignore that distinction.) What an individual does depends on the will of the team. I remember a game that we played against Slippery Rock University during my senior year. I scored 26 points in the first half but only two in the second half. I didn't go cold. My shooting percentage was the same in the second half as in the first half. The reason for the difference is that the team decided that I had scored enough points in the first half. They were correct since we won the game 77–67.

One aspect of Wilt's record that I haven't seen mentioned is that there was no three-point line when Wilt was playing. The only three-point plays in those days were a made shot and free throw. There should be an asterisk beside today's scoring records. On January 22, 2006, Kobe Bryant, who plays for the Los Angeles Lakers, scored 81 points against the Toronto Raptors to move into second place for NBA single-game scoring. (Both Wilt and Kobe were born and raised in the Philadelphia area.) If Bryant had played in 1962, he would have only scored 74 points and Wilt would still be in second place. For my money, the greatest single-game scoring effort in the NBA was Jordan's 63 in game 2 of the 1986 NBA playoffs versus the Boston Celtics. That's an NBA playoff scoring record for a single game that still stands.

Shooting and scoring are the fun parts of the game. My teammates will confirm that I had a lot of fun playing the game. They predicted that I was going to be a physicist and study magnetism because they said that the ball always seemed be attracted to iron after they passed it to me. They definitely won't be surprised to learn that most of this book is about shooting. In this chapter we analyze the flight of a basketball after it is released as a shot. In chapter 3, we consider a basketball that gets nothing but net and we analyze the mechanics of the shot itself. In chapter 4, we deal with a basketball that goes in after bouncing off the rim or backboard.

To start, let's pretend that the ball has eyes as in the cartoon in figure 2.1. We place the ball above the level of the hoop and a meter or so (a few feet) from the hoop. Also, we've labeled the angle below the horizontal from the ball to the middle of the hoop the angle to hoop as shown in figure 2.1.

We're interested in what the basketball sees when it looks at the hoop. That is a first approximation of how big the target is, the target being the portion of the hoop that the ball "thinks" that it can go through. Sketches of what the ball sees for different angles to hoop are shown in figure 2.2.

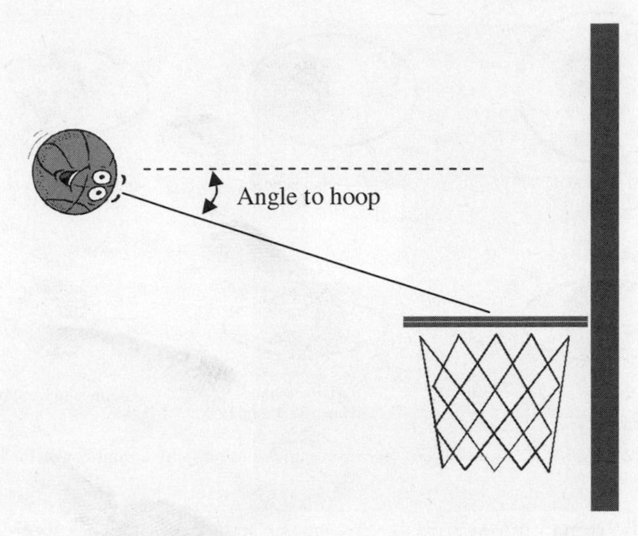

Figure 2.1. Definition of angle to hoop

Some of the drawings are similar to those in *The Biomechanics of Sports Techniques* by James G. Hay.[2]

The first and last sketches in figure 2.2 are included mainly for clarity. The first picture shows an angle to hoop of 90°. That picture represents what the ball sees if it drops straight down from above the basket. That could be achieved via a slam dunk where the ball starts directly above the hoop but is impossible to achieve for a shot starting anywhere else since it would take an infinite launch speed. An angle to hoop of 90° represents the maximum area (size of the target) that the ball could see when it looks at a hoop. The ball has the highest probability of going through the hoop for an angle to hoop of 90°. The last picture in figure 2.2 shows the situation for a ball with an angle of hoop of 0°. Clearly, an angle of hoop of 0° is not good if we want the ball to go through the hoop. In this case there is zero area that it can go through and the ball would collide with the front of the hoop if it proceeded on a straight line path.

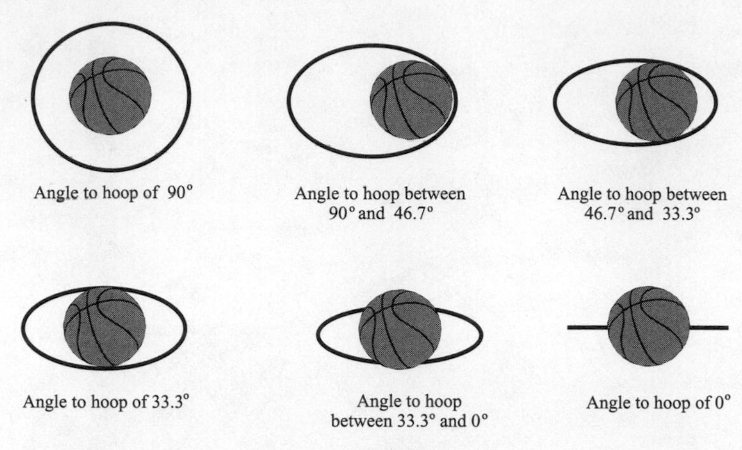

Figure 2.2. What a basketball sees for various angles to hoop for a men's basketball

The remaining pictures are what the ball sees for other angles to hoop. The important trend is that as the angle to hoop decreases from 90°, the area that a ball sees decreases. There are no subtleties for angles to hoop between 90° and 46.7° for a men's basketball. The lower angle is 45.7° for a women's basketball. For these ranges of angles, if the ball travels along a straight line, it is capable of passing through any part of the hoop. Also, the probability that the ball passes through the hoop decreases as the angle decreases. However, for angles below 46.7° for a men's basketball or 45.7° for a women's basketball, part of the area becomes excluded. If we look carefully at the sketch in figure 2.2 labeled angle to hoop between 46.7° and 33.3°, we see that a men's basketball cannot fit into the area to the right of the ball inside the rim. The same occurs for an angle to hoop between 45.7° and 32.1° for a women's basketball. This is important since the existence of excluded area makes it less likely that the ball will pass through the hoop.

For angles to hoop of less than 33.3° for a men's basketball and 32.1° for a women's basketball, the basketball cannot fit through any part of the hoop. For a men's basketball, see the sketch in figure 2.2 labeled angle to hoop between 33.3° and 0°. The cutoff angle to hoop of 33.3° for a men's basketball is slightly larger than the equivalent special angles quoted by Hay

(32.7°) and Brancazio (32°).[3] The difference is caused by the use of slightly different size basketballs and hoops in the calculations.

This discussion shows why a shot should have a large angle to hoop. The way to achieve that is to put a high arc on the trajectory. A lack of adequate arc is one of the biggest contributors to poor shooting. The problem is that low-arc shots flaunt the laws of physics. I have known this for many years and have used it to my advantage. For example, over the years, I have been very successful in shooting games against one particular friend. Since it's now unlikely that we'll play again, it's probably time to point out to Eric that he would have kicked my butt if he had just increased the angle to hoop for his shot by adding 0.3 m or so (a foot or so) to the height of his shot.

Although it is informative, the discussion in this chapter so far, angle to hoop and all that, is a spherical chicken as regards the flight of a basketball. The reason is that a shot basketball does not usually travel in a straight line. For simplicity, we'll first let only gravity act on the basketball. In that case a basketball shot at an angle to hoop other than 90° would proceed in a straight line to and through the hoop only if its speed were infinite. Even MJ in his prime couldn't shoot a basketball with infinite speed. What happens during a shot is that, for any angle to hoop other than 90°, the angle to hoop is constantly changing, that is, the basketball follows a curved path. The curved path occurs because no matter what direction the basketball is traveling, gravity always pulls it downward. This is the standard projectile motion problem that is usually studied in great detail in a typical general physics course. Those courses usually show that the curved path is parabolic, etc. However, in chapter 1 we saw that three other forces on a basketball traveling through the air, air drag, the buoyant force, and the Magnus force, can also be important. Later in this chapter, we'll include them in the model. However, for simplicity we'll start our analysis of the flight of the ball where the only force is gravity.

To proceed, we need to define a couple of new quantities. First, we define a different angle, the angle of approach. The *angle of approach* is defined as the *angle between the horizontal and the velocity of the ball when the ball is directly above the front of the hoop*. We will also specify how high the bottom of the ball is above the top of the front of the hoop. We'll call that *h*. By "front" we mean the point on the hoop closest to the shooter and not necessarily the point on the hoop closest to the top of the key.[4] Both the angle of approach and *h* are shown in figure 2.3.

While the top of the front of the hoop is useful mathematically, it also may be the most important place on the court for a shooter. Some coaches agree with that because they teach players to focus on the front of the hoop when they shoot. That has a lot of merit, because if a shooter can send the ball to a point a little above the top of the front of the rim it will go in most of the time. When I was at the foul line during a game, to help clear the mind and focus I tried to think "right up over the rim." Another reason to focus on the front of the hoop is that it is about all that is the same from court to court. The backboard and its background and even the hoop itself are different for different basketball courts, so focusing on the front of the rim helps eliminate the differences in the backboards or distractions caused by fans behind the backboard. Also, since we defined front as the point on the hoop closest to the shooter, it helps eliminate differences in the view of the basket from different positions on the court. Finally, as we will see in chapter 4, the physics says that trying for "right up over the rim" is the right thing to do.

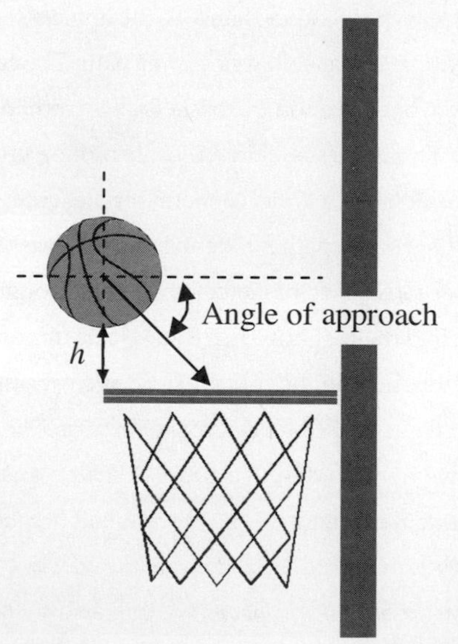

Figure 2.3. Representation of the angle of approach and *h*

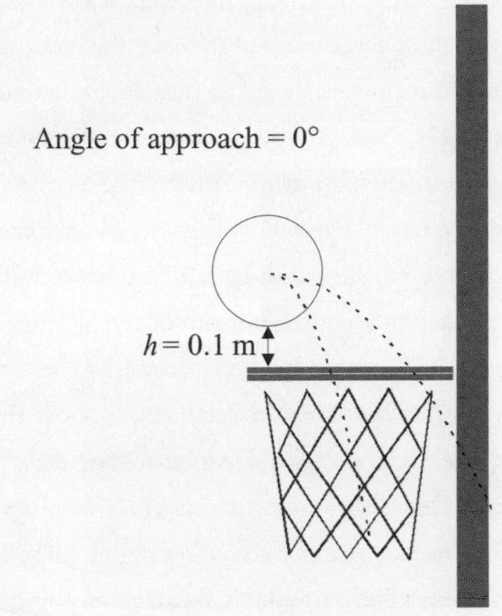

Figure 2.4. Extreme paths of the ball for an angle of approach of 0 degrees

We now consider how a basketball can go through the hoop and get nothing but net when the angle of approach is zero. An angle of approach of zero could be achieved by a finger roll or a shot from below and close to the basket. We also assume that the basketball has $h = 0.1$ m (4 in). Again, we are considering only the effect of gravity on the basketball. The extreme paths of such a shot are shown in figure 2.4.

Under these conditions, the ball gets nothing but net if the approach speed is between 0.62 m/s (1.4 mph) and 1.5 m/s (3.3 mph). These are the limits on the speed of release for the finger roll, for example. As h increases, the extreme speeds required for the ball to go through the hoop decrease since as the height increases, the ball takes a longer time to get to the hoop and thus would travel farther horizontally. The horizontal speed must decrease to compensate for this increased time. The same is true for the other angles of approach.

It is tempting to think that all or at least a very wide range of angles, speeds and heights can be used when a basketball is shot. However, the laws of physics put severe restrictions on the angle of approach, approach speed, and h that a basketball can have. What happens is that, for all basketballs shot from the same point in space so that they have the same angle of approach, there is an imaginary physics *window* that the basketballs must pass through to get nothing but net. A typical window is shown in figure 2.5. The bottom of the window occurs at the smallest value of h that can occur for a given angle of approach. The top of the window is at the largest value of h plus the diameter of the basketball. If the basketball passes through the imaginary window so that the bottom of the ball is below the bottom of the window, the ball hits the front of the rim. If the top of the ball passes through the imaginary window so that its top is above the top of the window, the ball hits the back of the rim or the backboard.

What is shown is only the vertical part of the window. The window also has horizontal dimensions since, for a given height, there is also a range of side-to-side positions where a basketball can get nothing but net. Since we'll learn enough by just considering the vertical part, we won't consider the horizontal aspects of the window in this book.

I made some calculations of the vertical size of the window for shots from a distance of 4.2 m (13.8 ft) from the front of the hoop (close to the foul

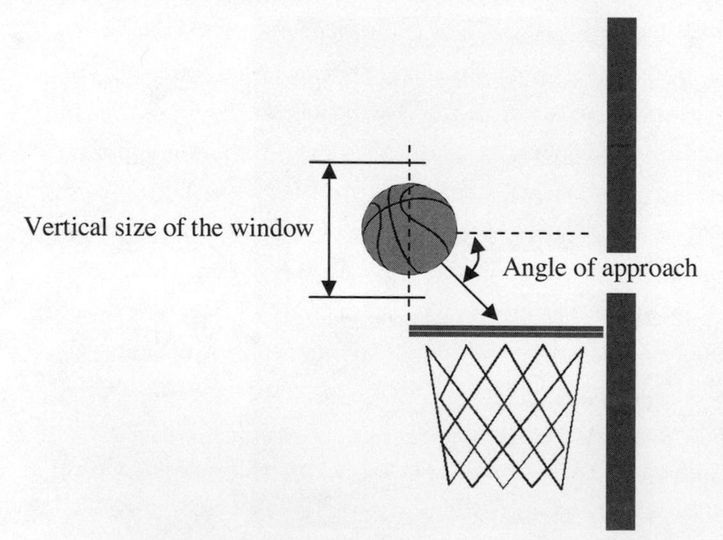

Figure 2.5. Definition of the window

line) and a height of release of 2.4 m (7.9 ft). A typical setup used for the calculations is shown in figure A.1 of appendix IV. Again, we are allowing only gravity to exert a force on the ball. The results of the calculations up to 56° are shown in figure 2.6. The calculations were arbitrarily cut off at 56° since higher angles of approach probably aren't of interest to most shooters. What is plotted is the size of the window minus the diameter of the basketball (this is equal to $h_{max} - h_{min}$) versus the angle of approach. This is plotted because it makes it easy to see the condition where it is not possible for the basketball to get nothing but net. That happens when a line crosses the x axis, that is, when the value is zero. At that angle, the size of the window equals the diameter of the basketball; so, the basketball just fits through the window. For smaller angles of approach, the size of the window is less than the diameter of the basketball. For those angles, the basketball is too large to fit through the window. Figure 2.6 shows that for angles of approach smaller than 29.0° for men and 27.9° for women it's impossible to get nothing but net for a shot from a distance of 4.2 m and height of 2.4 m. As expected, there is nothing special about the angles to hoop of 33.3° for a men's basketball and 32.1° for a women's basketball. Those cutoff angles occurred for straight line motion of a basketball through a hoop. Gravity makes the game a little easier since it increases the range of angles where the ball will get nothing but net.

Figure 2.6 also shows that the vertical size of the window increases as the angle of approach increases. The increase in the size of the window as the angle of approach increases is consistent with the qualitative model given at the beginning of this chapter where it was pointed out that a basketball sees a larger area as the angle to hoop increases. Our new, slightly more refined model gives a different (and more valid) reason that a basketball shot should have a reasonably high arc. Larger angles of approach have larger windows and that makes it easier to get nothing but net.

This explains something that I often wondered about. When shooting around, I found that it was fairly easy to shoot the ball through the hoop with an exaggerated (high) arc. In fact, that became part of my pregame ritual. I made sure that the first couple of shots during warmup had an abnormally high arc. That was a reminder to try not to shoot with too flat an arc during

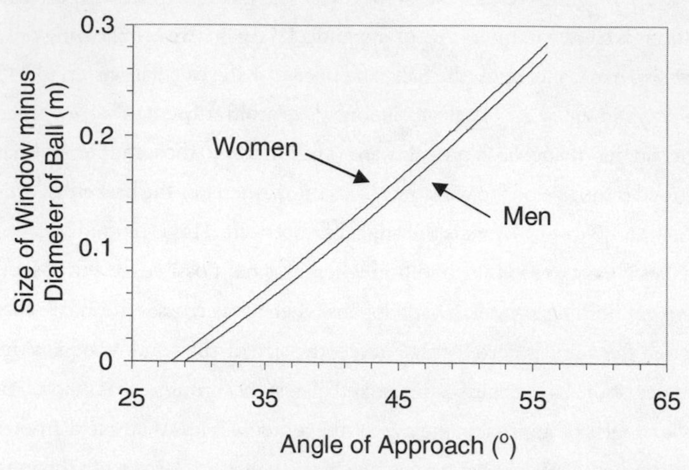

Figure 2.6. Plot of the size of the window minus the diameter of a basketball versus the angle of approach for a shot from a distance of 4.2 m and a height of 2.4 m, assuming that only the force of gravity acts on a basketball

the game. What was surprising was the large number of shots with an abnormally high arc that went through the hoop. It's probable that making a large fraction of those shots during the warmup was also helpful psychologically.

It follows from the discussion so far that players should shoot with as high an arc as possible. In general, they don't, though they sometimes do use an abnormally high arc when Shaq is around. One reason that players don't shoot with an extremely high arc is that the required forces are too large. For very large angles of approach, the arc must be very high (large launch angle) and thus the launch speed must be very large. Consequently, it requires a great deal of effort on the part of the shooter. What is subtle is that, for very small angles of approach, the launch speed must also be very large. That is because very small angles of approach require very small launch angles and it takes a lot of initial speed to shoot a basketball a great distance if it is launched at a very low angle. The net result is that somewhere between large and small launch angles there is a minimum launch speed for the ball. Brancazio has suggested that the minimum launch speed is what good shooters strive for.[5] He also shows how to calculate the launch angle for the minimum launch speed when the basketball is under the influence of only gravity.[6] We could easily reproduce that calculation. However, we know that forces other than gravity can be important. Consequently, before we make further use of a model that includes only the force of gravity, let's evaluate just how good it is.

I used the digital camcorder described in chapter 1 to record a shooting session with one of my ex-students and former Navy star, 25-year-old Mike Heary (fourth on the Navy all-time list with 1,590 points and second in career free-throw percentage at 83.6%) and a 58.8-year-old (1,303 points with free-throw and field-goal percentages of 86.9% and 51.0%[7] at Westminster College, New Wilmington, PA). Okay. I am the 58.8 year old. The data were treated similarly to those in figure 1.3 in chapter 1.

We start by analyzing one of Mike Heary's free throws. Free-throw shooting (also called foul shooting) is an important part of the game. Huge numbers of games are won and lost because of free-throw shooting. It may surprise you to learn that the NCAA collegiate career record for free-throw shooting percentage is held by Andy Enfield of Division III Johns Hopkins University who compiled an astounding career free-throw shooting percentage of 92.5% having made 431 out of 466 over 108 games.[8] Enfield played from 1988 until 1991. That record has survived a recent, serious challenge from J. J. Redick of Duke University. During the 2004 season Redick had the best free-throw percentage ever for a sophomore, 95.3%. That helped give him a percentage going into his senior year of 93.9%. Redick had only an "average" senior year and ended his career at 91.2%, good enough for fourth place on the all-time list. Nonetheless, Redick is one the best pure shooters to play the game in a long time. Are women better free-throw shooters than men? The jury is still out on that question though the evidence points to an answer of probably not yet. The top two women free-throw shooters, Kandi Brown of Morehead State University and Brooke Lassiter of Louisiana Tech University both have career free-throw percentages of 91.5%.[9] Brown holds the record having made 357 out of 390. Their percentages are larger than those for the two men on the list after Enfield, Gary Buchanan of Villanova University and Korey Coon of Illinois Wesleyan (both 91.3%). However, the percentages for the women leaders lower on the list are all smaller than those for the corresponding men.

The position of the ball every 1/60 s for one of Mike's foul shots is shown by the open circles in figure 2.7. As usual, Mike's foul shot gets nothing but net. The launch speed and launch angle were determined from the data. The theoretical path of a basketball launched at the same speed and angle as Mike's shot was calculated assuming only the effect of gravity. This is the classic projectile motion problem discussed in most general physics courses. The theoretical path is shown by the solid line in figure 2.7. The theory incorrectly predicts that the shot misses. According to the theory, the ball overshoots the basket and, at best, hits the back of the hoop.

The failure of the theory is not surprising considering the discussion in chapter 1. Fortunately, that discussion gives us some ideas about how to improve the model. For example, in chapter 1 I pointed out that at the typical speeds involved in a basketball game, air resistance is about 10% of the force of gravity. We expect air resistance to help improve the agreement between experiment and theory since figure 2.7 shows us that a basketball under the influence of only gravity overshoots the basket. Since air resistance slows the ball, its inclusion should reduce the overshoot, so that is the first thing that we will check out. I used the techniques described in appendix III to cal-

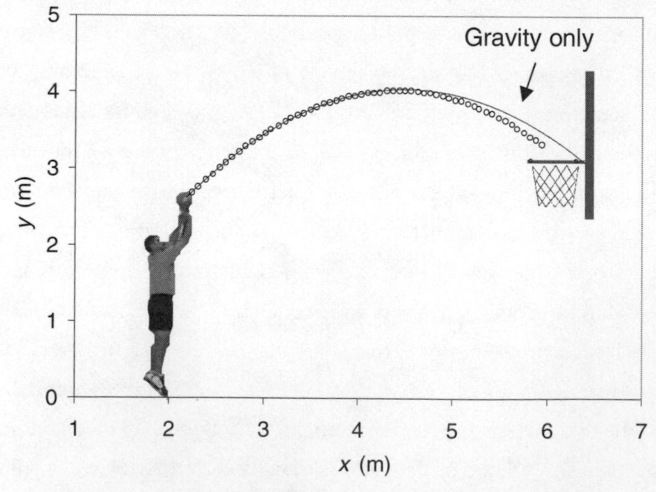

Figure 2.7. Plot of the experimental path of a basketball during a foul shot by Mike Heary and the theoretical path based only on gravity

culate the theoretical path of Mike's foul shot if it is under the influence of both gravity and air resistance. The theoretical path is shown by the line in figure 2.8. Again, the circles represent the actual position of the ball every 1/60 s.

Air resistance has the expected effect since the new theoretical path is lower than the theoretical path based on gravity alone. However, the new theoretical path is now lower than the actual path of the ball because air resistance overcorrects. It is also apparent from figure 2.8 that there is not much difference between the theoretical and actual paths of the ball. In fact, the ball still gets nothing but net. Consequently, from a basketball point of view, we could stop there. Certainly, any coach or player would be happy with the results of the theoretical foul shot. From a physics point of view, though, the theory is not satisfactory. If we really want to be able to predict things about shooting a basketball, the physics needs to be better.

It is apparent from figure 2.8 that some lift is needed to increase the height of the theoretical curve slightly. Some of the missing lift is provided by the Magnus force because properly shot basketballs have backspin. Several reasons why properly shot basketballs should have backspin are discussed in the remaining chapters. To see why shot basketballs with backspin experience lift, it might be useful to look at figure 1.1 where the ball is moving downward and spinning counterclockwise. In that case, the Magnus force

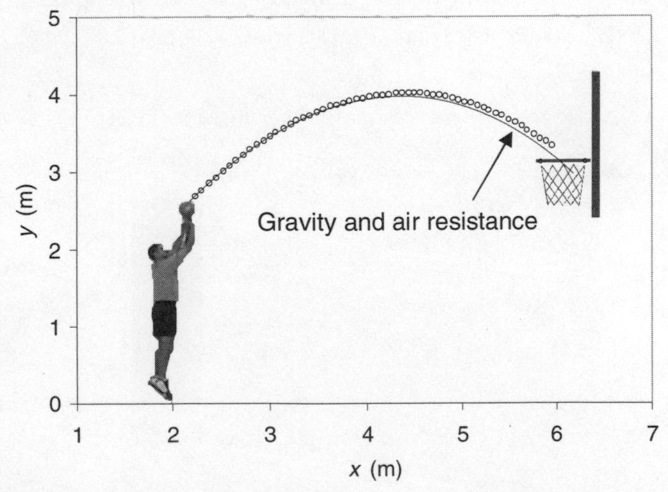

Figure 2.8. Plot of the experimental path of a basketball during a foul shot by Mike Heary and the theoretical path based on gravity and air resistance

is to the right. If we rotate the book by 90° so that the ball appears to be moving to the right, the Magnus force will be pointing up, hence the lift. While rotating the book by 90° shows the correct direction of the Magnus force for that situation, the direction of the buoyant force and gravity would be wrong so it might be better to look at figure 2.9 where all of the forces are drawn correctly for two different velocities of a basketball spinning counterclockwise. In the sketch on the left side of figure 2.9, the velocity of the basketball is up and to the right. In this case, the Magnus force is up and to the left as shown. When the velocity of the basketball is down and to the right, the Magnus force is up and to the right as shown in the sketch on the right side of figure 2.9. This is a true lift force since it is in the same direction as the lift force on an airplane wing if the airplane were traveling in the same direction as the basketballs shown in figure 2.9. (This assumes that the airplanes are right side up, that is, not traveling upside down.)

To calculate the Magnus force, it is necessary to know the rotational speed of the basketball. To determine the rotational speed, a strip of duct tape was added to the basketball and close-up videos of the ball were taken during some shots. The times at which the ball had rotated through increments of one-fourth of a revolution were determined and the angle of rotation was plotted versus time. The slope of the graph gave the rotational speed in revolutions per second. The number of shots analyzed is given in parentheses and the average values of the speed of rotation are given in the table 2.1.

Table 2.1. *Speed of rotation of a basketball in revolutions per second*

	Jump shots from about 2 m	Jump shots from the foul line	Free throws	Three-point jump shots
The author	2.04 ± 0.07 (15)	2.19 ± 0.08 (21)	2.19 ± 0.10 (14)	2.28 ± 0.10 (12)
Mike Heary		1.86 ± 0.10 (6)		2.03 ± 0.09 (5)

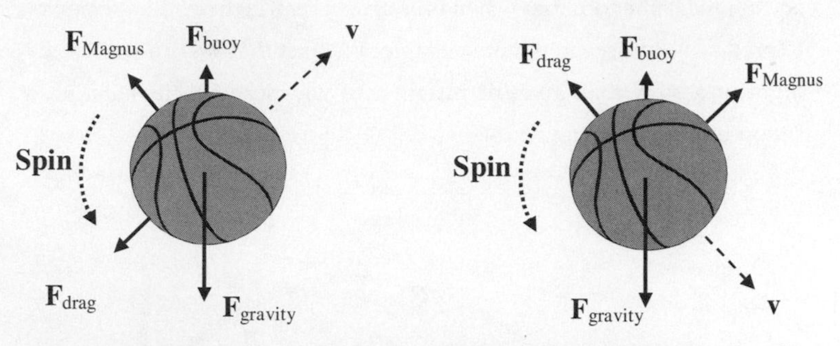

Figure 2.9. Forces on a basketball traveling up and to the right and down and to the right

The general result is that both shooters put a rotation on the ball of about two revolutions per second. It also appears that the greater the distance of the shot, the greater the speed of rotation. This makes sense. Basketballs shot from a greater distance usually have a greater (linear) launch speed and it is "natural" to put more spin on the ball when it is launched with a greater linear speed. The reasons for this are discussed in chapter 3.

The Magnus force was added to the model using the rate of spin for Mike Heary's jump shots from the foul line. The results of the calculation are shown by the line in figure 2.10. Including the Magnus force improves the agreement between the theory and the experimental results but there is still a small difference. The cause of the remaining difference is the other force that was discussed in chapter 1—the buoyant force. The buoyant force is usually ignored in projectile motion problems. That is appropriate so long as the object is small and its weight is large. However, a basketball is relatively large for its weight. In chapter 1, it was pointed out that the buoyant force on a basketball is about 1.5% of its weight. I added the buoyant force to the model and the complete four-force (gravity, air resistance, Magnus force, and buoyant force) model resulted in the line shown in figure 2.11.

There is good agreement between the complete four-force model and the experimental data. Consequently, it appears that this model enables us to predict the flight of a shot basketball. To me, this is an example of the ultimate goal (not a pun) of any exercise in physics. What physicists try to do is develop theories and write equations for measur-

able physical phenomena. When they are successful they have the power to predict. As we see in the next chapter, the fact that we can predict the flight of a basketball gives us insight into why some players are good shooters and others aren't.

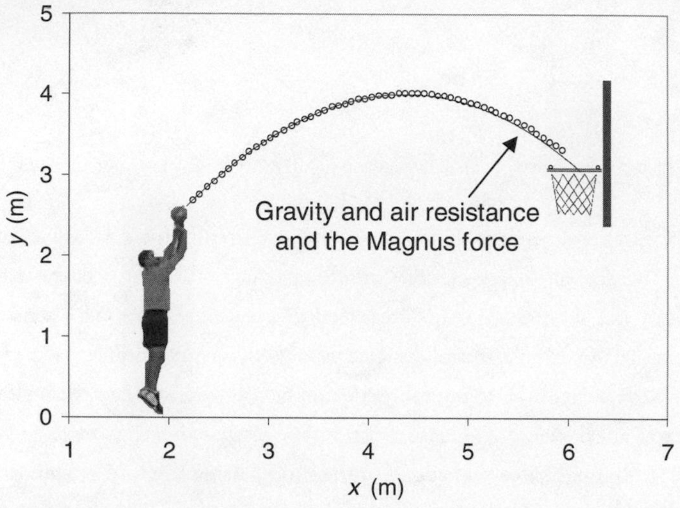

Figure 2.10. Plot of the experimental path of a basketball during a foul shot by Mike Heary and the theoretical path based on gravity, air resistance, and the Magnus force

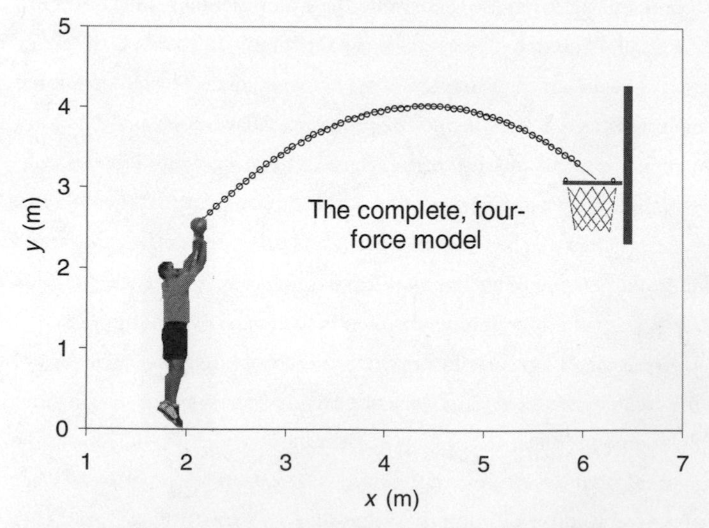

Figure 2.11. Plot of the experimental path of a basketball during a foul shot by Mike Heary and the theoretical path based on the complete four-force model

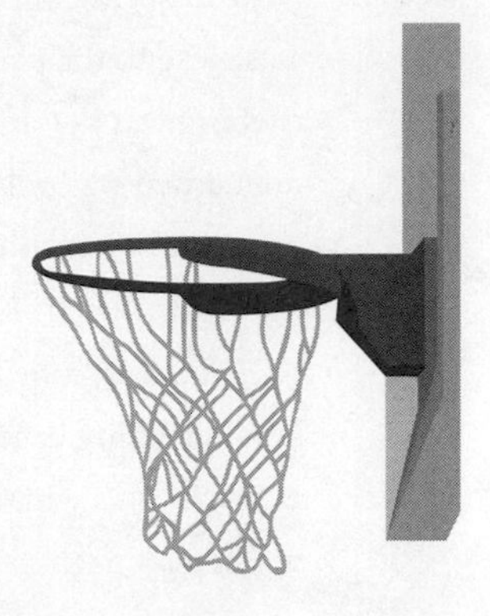

Nothing But Net

What is it that a good shooter does? That's the question that we try to answer in this chapter and the next. When I think of a single clutch shot and nothing but net, one game comes to mind. It's probably not the one that you think. It's not Christian Laettner's shot that won the NCAA East Regional championship game for Duke over the University of Kentucky on March 28, 1992. I am convinced that Laettner's shot grazed the rim. We'll revisit Laettner's shot in chapter 4. The game that I am thinking of is the NCAA championship game between the University of North Carolina and Georgetown University on March 29, 1982. My wife will confirm that I watched that game because of my reaction to Fred Brown's pass.[1] UNC had a 63–62 lead with 15 seconds and Brown was bringing the ball up the court for Georgetown. He stopped, looked around, then inadvertently passed the ball directly to James Worthy[2] who was playing for UNC. Worthy ran some time off the clock until he was fouled. He missed two free throws but it was too late for Georgetown who only had time for an outlet pass after the second missed foul shot and a desperation shot.

Too much emphasis is given to the mistake by Fred Brown. If he hadn't committed that basketball faux pas, Georgetown would have only had a chance to win. There is no guarantee that they would have. What really lost the game for Georgetown is that when they had the lead at 62–61 with 15 seconds remaining they left a UNC player unguarded. No one was close to the player when he made the pretty, 16-foot jumper that got nothing but net and won the game for UNC. The player who was left unguarded was only a freshman but his name was Michael Jordan. Jordan has said, "That shot put me on the basketball map."[3] The shot also put the UNC coach, Dean Smith, on the basketball map. It was his first NCAA championship. He would go on to win another title 11 years later before retiring in 1997 with an NCAA record 879 victories in 36 seasons. Dean Smith's win total was eclipsed on March 22, 2005, by Pat Summitt, women's coach at the University of Tennessee. Coach Summitt has received every award imaginable both as a player and a coach. She has won six NCAA championships and gives no indication of slowing down. The only remaining question, it would seem, is whether she will equal or exceed the ten NCAA titles that John Wooden won while coaching at UCLA. Even though he coached great players such as Bill Walton, Lew Alcindor (Kareem Abdul Jabbar), Walt Hazzard, Gail Goodrich, Lucius Allen, Mike Warren, Sidney Wicks, Curtis Rowe, Henry Bibby, Keith Wilkes, Richard Washington, and Dave Meyers, winning was never automatic. His coaching genius led to seven straight NCAA titles. My guess is that that record will never be broken.

To start our analysis of what a good shooter does, let's use the full four-force model to recalculate the vertical size of the window that a basketball must pass through if it is to get nothing but net. We will spend the most time analyzing foul shots because the point of release is fairly well defined. It is fairly well defined because the feet are usually fixed at the foul line. We will let the point of release be a horizontal distance from the backboard of 4.2 m (14 ft) and a height of 2.4 m (8 ft) above the floor. The results of the calculations based on the full four-force model are shown in figure 3.1. The vertical axis in figure 3.1 is different from figure 2.6 in that the actual maximum and minimum heights above the rim for the ball are shown in figure 3.1. A plot similar to figure 2.6 would be obtained if the difference between the upper and lower curves in figure 3.1 minus the diameter of the ball were plotted versus the angle of approach.

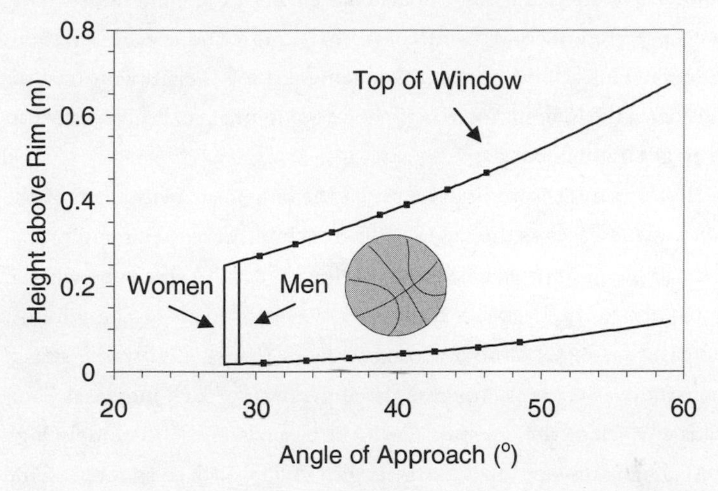

Figure 3.1. Plot of height above the rim of the top and bottom of the window for foul shots

The ball cannot get nothing but net if the angle of approach is less than 28.8° for men and 27.8° for women. Those limits are shown by the vertical lines in figure 3.1. To the left of the vertical lines, the diameter of a basketball is greater than the size of the window. It will bounce off the rim for angles smaller than 28.8° and 27.8°. These angles are slightly smaller than the angles of 29.0° and 27.9° calculated using only the force of gravity. This implies that the other three forces, air resistance, the Magnus force, and the buoyant force make it slightly easier to get nothing but net.

The basketball in figure 3.1 is shown at an angle of approach of about 40°. The reason for the choice of 40° as an example will become apparent later in this chapter. At this angle of approach the size of the window for a men's basketball is about 0.340 m (about 13.4 in). The size of the window for a women's basketball is about 1% larger. These dimensions are to be compared with the diameter of a women's basketball (about 0.23 m) or a men's basketball (about 0.24 m). The extra vertical height (vertical margin for error) for a basketball is only 0.11 or 0.10 m (about 4 in). This is not a lot and shows why not everyone is a good shooter. For comparison, I calculated the size of the window assuming that only gravity controls the basketball and found a value of about 0.338 m. This also implies that, in general, the other three forces, air resistance, the Magnus force, and the buoyant force, make it slightly easier to get nothing but net.

It is apparent from figure 3.1 that the height of the bottom of the window increases as the angle of approach increases. This shows that if the angle of approach is larger the lowest that a ball can be when it is just above the basket is higher. However, the top of the window increases faster than the bottom increases. Consequently, the size of the window becomes larger as the angle of approach increases. This leads us back to the question "Why don't players use extremely high arcs?" High arcs would give large angles of approach so shooters could take advantage of large windows. Because we now have a better model of the flight of a basketball, we can get some insight into the answer to this question.

In an attempt to answer this question I digitized several foul shots by Mike Heary and me. The open circles shown in figures. 2.7, 2.8, 2.10, and 2.11 represent one of the sets of data. I next used Videopoint® to calculate the speed and horizontal and vertical components of the velocity at each of the points. From the velocity I calculated the angle of the velocity above the horizontal at each of the points. For each of the shots, the time of launch was also determined. That was estimated from the video as the time when the ball left the hand. The launch angle, launch speed, and launch position were then determined from the data as the angle, speed, and position of the basketball at the launch time. Because there was scatter in the data for the angle and speed versus time, both cubic and quartic fits to the data versus time were made to obtain smoothed representations of the data after the launch. Those fits were used to determine the best values for launch angle, launch speed, and launch position. The average results are listed in table 3.1. The launch position is tabulated as a horizontal distance from the backboard, d_{launch}, and height above the floor, h_{launch}.

Table 3.1. *Average launch speed, launch angle, and launch position for foul shots*

	θ_{launch} (°)	v_{launch} (m/s)	d_{launch} (m)	h_{launch} (m)	Comment
Mike Heary	51.5 ± 1.6	6.78 ± 0.25 (15.2 mph)	4.3 ± 0.1 (14.1 ft)	2.62 ± 0.02 (8.6 ft)	5 made
The author	51.0 ± 1.0	6.99 ± 0.07 (15.6 mph)	4.2 ± 0.1 (13.8 ft)	2.43 ± 0.01 (8.0 ft)	5 made
The author	49.1	7.2	4.2	2.45	long

There is a difference in the way that the data were selected for analysis. For me, only foul shots that were clearly nothing but net were analyzed, while for Mike Heary a full range of made shots was analyzed. That included both nothing but net shots and those just over the front of the rim and bouncing off the back of the rim. That explains the smaller standard deviation for my foul shots and has nothing to do with the quality of the shooter. Mike is the better shooter, at least now.

What is most striking about the data is that the launch angles for my foul shots and for Mike Heary's are both about 51°. This suggests that there is something that the techniques used by both of these shooters (okay—one shooter and one imposter) have in common. As pointed out in chapter 2, it has been suggested that players shoot at the minimum launch speed.[4] To test this suggestion, the launch speed for various launch angles was recalculated using the complete four-force model and the average launch point of my foul shots. The results are shown by the triangles in figure 3.2. The dashed line is a best-fit quadratic that was used to help estimate the position of the minimum launch speed. The launch angle where the launch speed is a minimum is 48.2° is about 3° smaller than the measured launch angle of about 51°. It would appear that a difference exists between the theory and the experimental results.

A few possible reasons for the difference are as follows. It could be that the shooters are just shooting at a higher angle (51° versus 48°) to take advantage of a larger window. Also, I hate to admit it, but I can't guarantee that the 3° difference is real. 3° is a small angle and, as with any experiment, there is uncertainty. The uncertainties quoted in table 3.1 are based only on the scatter in the results. There may be some systematic error that I have overlooked. However, the 3° difference has another explanation that is interesting. I used the full four-force model to calculate the approach speed (speed just above the front of the hoop) versus launch angle for basketballs that go through the center of the top of the hoop. The results for the approach speed versus launch angle are shown by the squares in figure 3.2.

The solid line is a best-fit quadratic that was used to help estimate the launch angle that gives a minimum approach speed. I was intrigued to find that the launch angle that gives a minimum approach speed is about 50.8°. This is close to the launch angle that is measured for foul shots by both shooters and suggests the following "hoopothesis": What a good shooter does is minimize the approach speed. This makes sense since a minimum approach speed will give the "softest" shot. That is because the slowest approach speed (speed just above the front of the hoop) will provide the weakest collision or interaction of the basketball with the net. It is easy for a shooter to judge "softness" based on the way that the ball hits the rim for a slightly errant shot. The classic "unsoft" shot even has a name—a "brick." My guess is that what a conscientious shooter does is study, probably subconsciously in most cases, how slightly errant shots hit the rim. The especially observant shooter probably can judge the "softness" of a shot by the way that the basketball interacts with the net. The shooter then makes adjustments in the shot until the softest shot is achieved.

Armed with this information, I tried to determine whether physics could explain why tall players are often poor foul shooters. It has already been mentioned that Wilt Chamberlain was notorious for poor foul shooting. His career NBA foul-shooting percentage was 51.1%. These days, it's Shaq. As of the writing of this book, his career foul-shooting percentage is 53.1%. I made some calculations for foul shots launched from a greater height, 2.75 m (9 ft) versus 2.46 m (8 ft). I found that, at launch angles up to 56°, which is where the calculations stopped, the size of the

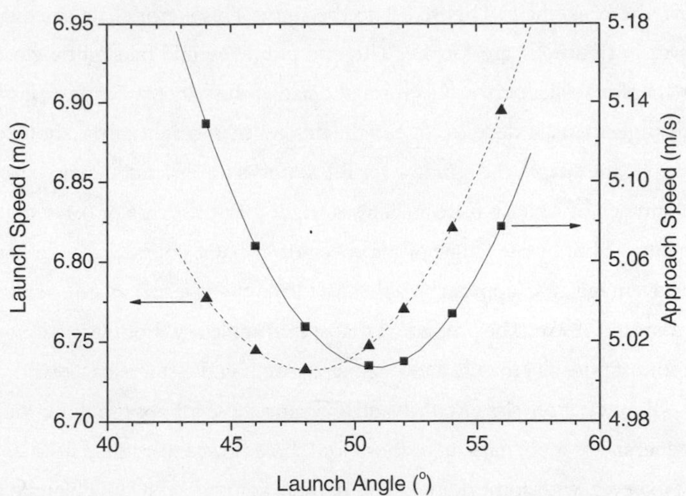

Figure 3.2. Plot of launch speed and approach speed versus launch angle for foul shots

window is about 0.002 to 0.005 m smaller for the tall person. For example, at an angle of approach of 40° the size of the window for a tall person is 0.336 m versus 0.340 m for a shorter person (me). The difference is not large, but it can be a large percentage of the vertical margin for error (height of the window minus the diameter of the basketball). At an angle of approach of 40°, the vertical margin for error is 5% smaller for the tall person. That percentage increases as the angle of approach or the launch angle decreases. In addition, the minimum possible angle for the basketball to get nothing but net is about 0.2° larger. Consequently, the tall person has a slightly smaller range of angles available to get nothing but net. This shows that, in general, it is slightly easier to shoot a basketball from a height of 2.46 m (9 ft) than from 2.75 m (8 ft).

If a tall person works to achieve the softest shot, however, she or he has an advantage. For a tall person, the launch angle that achieves the minimum approach speed is about 48.7°. This angle is lower than the value of 50.8° for me. This shows that a taller person should shoot at a lower angle. Further, the launch angle of 48.7° has a corresponding angle of approach of about 42.8° compared with 40° for me. This larger angle of approach results in the basketball having access to a larger vertical window. The way to see this is to move the basketball in figure 3.1 to the right. This is not exact since the curves in figure 3.1 are for me. The two curves would be slightly closer together for a taller person. However, the size of the window increases faster with angle than the decrease due to the difference in height of the shot. For example, the size of the window for the softest shot of a tall person (angle of approach of 42.8°) is 0.365 m. This is larger than the value of 0.340 m for the softest shot by me (angle of approach of 40°). Of course, I would have the advantage at an approach angle of 42.8° in that the size of my vertical window is 0.369 m. The problem is that at that angle my shots are no longer the softest possible, that is, they begin approaching the status of a brick.

The laws of physics give Wilt and Shaq an excuse for poor foul shooting. In general, it is more difficult to shoot foul shots from 2.75 m than from 2.46 m. However, with some adjustment, namely adjusting the launch angle to lower values, they should be able to achieve an even higher percentage. Another seven-footer, Dirk Nowitski who plays for the Dallas Mavericks, shows that a tall person can be a good foul shooter. As of December 2005, Nowitski's NBA career foul-shooting percentage was 85.6%.

To further test the hoopothesis that what good shooters do is minimize the approach speed (shoot the softest shot), some (almost) three pointers were analyzed and the results are given in table 3.2.

The results show that both shooters shoot at a lower angle from a greater distance. Using the average launch position for Mike Heary's

Table 3.2. *Average launch speed, launch angle, and launch position for (almost) three-point shots*

	θ_{launch}(°)	v_{launch}(m/s)	d_{launch}(m)	h_{launch}(m)
Mike Heary	45.6 ± 1.0	8.1 ± 0.1 (18.1 mph)	6.1 ± 0.1 (20.1 ft)	2.80 ± 0.04 (9.2 ft)
The author	45.2 ± 0.8	7.91 ± 0.09 (17.7 mph)	5.9 ± 0.1 (19.3 ft)	2.62 ± 0.02 (8.6 ft)

shots, the launch angle for the minimum launch speed was calculated to be 44.5°. This is again lower than is observed, and the calculated launch angle for minimum approach speed, 45.8°, is once again in better agreement with the data.

Finally, data for some jump shots from just beyond the foul line are given in table 3.3. The observed launch angle and launch speed are between those observed for the foul shots and (almost) three pointers. The theoretical value for minimizing the approach speed, 48.2°, is once again closer to the experimental values than the theoretical value for minimizing the launch speed, 45.5°. It would appear, then, that all of the results are explained if what a good shooter does is to launch the basketball so as to minimize the approach speed.

From time to time in this chapter we have compared the size of the window for foul shots with the size of a basketball. This gave us an idea of the vertical margin for error or the control required for a shot to get nothing but net. Another way to characterize the control required for a person to be a good shooter is the range of launch speeds allowed for a basketball to get nothing but net. The launch speed corresponding to the launch angle of 50.8° which minimizes the approach speed for a foul shot using either a men's or women's basketball is 6.75 m/s. If the launch angle is fixed at 50.8° it is found that the maximum launch speed to get nothing but net using a women's basketball is slightly greater than 6.80 m/s

Table 3.3. *Average launch speed, launch angle, and launch position for shots from just beyond the foul line*

	θ_{launch}(°)	v_{launch}(m/s)	d_{launch}(m)	h_{launch}(m)
Mike Heary	49.6 ± 0.9	7.2 ± 0.2 (16.1 mph)	4.7 ± 0.2 (15.5 ft)	2.76 ± 0.03 (9.0 ft)
The author	48.5 ± 2.0	6.86 ± 0.06 (15.3 mph)	4.6 ± 0.1 (15.1 ft)	2.53 ± 0.03 (8.3 ft)

and the minimum is slightly less than 6.71 m/s. The setup and details of this calculation are given in appendix IV. Consequently, the speed must be controlled to within about 0.7%. Because a men's basketball is larger the requirements are slightly more stringent. The maximum launch speed is slightly greater than 6.79 m/s and the minimum is slightly larger than 6.71 m/s. This is another indication of why not everyone is a good shooter.

The reverse situation is also informative. Suppose that we fix the launch speed and vary the launch angle. If the speed is set at 6.75 m/s, it is not possible to hit the back of the hoop by varying the launch angle. What happens is that as the launch angle is decreased from 50.8° the trajectory begins to move toward the back of the rim then stops increasing as the launch angle approaches 48.4° for women and 48.3° for men. The trajectory then begins to move toward the front of the hoop and the ball strikes the front of the hoop for angles smaller than 44.8° for women and 45.0° for men. For very small launch angles, the ball goes beneath the hoop, missing the hoop completely. When this happens the fans for the opposing team usually begin chanting "Air ball, air ball . . ." When the launch angle is increased from 50.8° the trajectory immediately begins moving toward the front of the hoop. For launch angles less than 53.5° for women and 53.3° for men, the ball strikes the front of the hoop. For very large launch angles, the fans for the opposing team have another opportunity to practice their chant.

Basketball fans are a special group. They can be critical, vocal, and downright abusive. Just about the time that you give up on them, however, something special happens. Think, for instance, of what took place on the night of March 16, 1986. The Middies of the United States Naval Academy, led by the incomparable David Robinson, had already defeated Tulsa University in the first round of March Madness. They were now playing Syracuse University in the second round of the NCAA, and, to make matters worse, the game was at Syracuse. The situation is best described by the rhetorical question raised by a Syracuse newspaper: "Will the masses 30,000 strong turn out in full battle array, descend upon the great bubble at $17 a pop to watch the Orange play a bunch of shorthairs they beat by 22 (89–67) 98 days ago?"[5] The answer was not what the journalist expected. Robinson scored 35 points, grabbed 12 rebounds, blocked 7 shots, stole the ball 3 times, and Navy won 97–85. Middies coach Paul Evans summed up the team effort by saying: "Not bad for a bunch of shorthairs." As for Robinson's performance, the Annapolis *Evening Capital* said it best, "So dominant was Robinson that even the Syracuse fans, who don't like anyone in an enemy uniform, gave the Mid a standing ovation when he left the game on 5 personal fouls with 2:43 to play." Kudos to the Syracuse fans—you showed a lot of class on that night. Robinson's performance was a coming out. Until that game he had been ignored by the basketball world. Afterward, he was the talk of the airwaves and highlight films were shown everywhere. It was on that night that Brent Mussberger gave Robinson the nickname "The Admiral." The nickname stuck and Robinson went on to a brilliant 14-year NBA career, all with the San Antonio Spurs.

There are still other things that we can learn from a video analysis of a basketball shot. For example, it is usually straightforward to evaluate why a shooter misses. For a straight-on shot, one that is on a line to the center of the hoop, a shot misses because some combination of launch speed and launch angle is wrong. If the reason is one or the other, launch angle or launch speed, the reason for the miss is usually obvious from the video analysis. An example of this is given in table 3.1 for one of my foul shots that missed long. The launch speed was too large and the launch angle was too small for that shot.

This is of more than just academic interest. Quantitative video analysis of the flight of a shot basketball should be a useful training tool. Of course, today's coaches do qualitative video analysis. They look for bad basketball technique and tendency based on their experiences. At the present time, I do not know of any coaches who do a quantitative video analysis even though it is a straightforward, inexpensive, and powerful tool. It should be easy to identify trends in the cause of missed shots just from the flight of the ball. Drills could easily be devised to correct bad habits.

We have just scratched the surface of the use of video analysis since we have only studied the flight of the ball. We can also learn some useful things by analyzing the path of the ball while it is in contact with the hand. That is probably more important than analyzing the flight of the ball because it is the forces of the hand on the ball that give it its initial velocity and spin. The initial velocity and spin of the ball are everything since they (plus the four forces) completely determine the subsequent flight of the ball. Sometimes we forget that the hand has nothing further to do with the ball once it leaves the hand; a shooter only controls the ball during the short time that the ball is in contact with the hand. It is during that short time that small irregularities in the forces of the hand on the basketball occur that make the ball miss, that is, the short time of contact of the hand with the basketball makes or breaks the shot.

Many aspects of what happens while the ball is in contact with the hand are important. For example, the release point is important to both the shooter and the person on defense. One of my favorite techniques was to not use maximum extension of the arms for my average shot, so my release point was not at the maximum height achievable. What happens is that good defenders gradually identify the release point of the shooter, then occasionally go for the ball in an attempt to try to block the shot. I remember that at least once a game the person on defense would go for the ball during a shot, then wonder why he hadn't blocked it. The reason was that I always had a little height left so that when he went for the ball, I raised it a little so I was able to get the shot off.

Figure 3.3 shows pictures of Rick Barry, Michael Jordan, and Peja Stojakovic shooting. Barry is the only person ever to lead the NCAA, ABA, and NBA in scoring at various stages of a career and was known as a scoring machine when he was playing. His NBA field goal percentage is 44.9% over 794 games. In addition, he had a 3-point success rate of 33% and was a 90.0% career foul shooter. Jordan's career NBA field goal percentage, 49.7% over 1,072 games, is higher than Barry's. Even though slam dunks probably accounted for a larger portion of Jordan's shots, he was an excellent outside shooter.[6] His 3-point success rate, 32.7%, is close to Barry's though his foul-shooting percentage, 83.5%, is lower. Stojakovich had a 46.1% field goal percentage over his first 540 NBA games. His foul-shooting percentage, 89.3%, is almost as good as Barry's and his 3-point shooting percentage is an impressive 39.9% (1,121 of 2,031). These three players are outstanding shooters. The pictures in figure 3.3 were chosen as good examples of three stages of a jump shot. The picture of Barry shows proper positions for the hands, basketball, and body just after the jump. The picture of Jordan shows correct form close to the release point, which occurs at the top of the jump. The picture of Stojakovich shows the position of the hands after a proper follow-through. As will be discussed in detail later in the book, the position of Stojakovich's right hand is evidence of a proper wrist snap.

An important aspect of a shot is the total time that the ball is in contact with the hand, that is, the release time. It is important to try to minimize the release time. Players who keep the ball in contact with the hand a very short time are said to have a quick release. Needless to say, players with a quick release are tougher to defend. In my opinion they're also more fun to watch. I haven't analyzed any of his shots, but I am certain that Jason Williams has one of the quickest releases in the NBA.

We start our video analysis of the time in contact with the hand by analyzing a couple of jump shots from just beyond the foul line. The pictures associated with figures 3.4 and 3.5 show the position of the ball every 1/60 s during jump shots by Mike Heary and me. In the graphs, the open circles represent the vertical velocity of the ball versus time. The filled (solid) circles represent the vertical velocity of our heads versus time. Positive values indicate upward velocity and negative values indicate downward velocity. The vertical dashed line in each of the graphs is drawn at the estimated time of release of the ball, about 0.62 s in both cases.

The information that we are focusing on is to the left of the dashed line in each graph. During that time, the ball is in contact with the hand. Our shots have some features in common. According to the data (filled circles), our heads both move downward for about the first 0.25 s. During this time the knees bend, and so on, in preparation for the jump. Beginning at about 0.25 s there is a rapid upward acceleration that reaches about 20 m/s^2 (about 2 gs). This acceleration is caused by the initial upward push of the floor on

(a)

Figure 3.3. Pictures showing the different stages of a jump shot. (a) Rick Barry of the Houston Rockets prepares to take a shot against the New York Nets (undated). (b) Michael Jordan of the Washington Wizards prepares to the release the ball during a shot against the Toronto Raptors during an NBA game at Air Canada Centre on December 15, 2002 in Toronto, Canada. (c) Peja Stojakovic of the Indiana Pacers follows through after a shot against the Atlanta Hawks on February 24, 2006 at Conseco Fieldhouse in Indianapolis, Indiana. (a) ©2006 Ron Turenne / NBAE / Getty Images. (b) ©2006 Focus on Sport / NBAE / Getty Images. (c) ©2006 Ron Haskins / NBAE / Getty Images.

(b)

(c)

the feet. That is the N3L reaction force to the feet pushing on the floor. This is traceable to the straightening of the legs. There is a leveling of the upward velocity from about 0.3 s to about 0.45 s. That also requires an upward push of the floor on the feet. This force is traceable to the flexing of the ankles. What that force does is balance the downward force of gravity. This makes the acceleration zero and thus the velocity constant. At about 0.45 s, the shooters leave the floor and the velocity of the person begins to decrease (increase downward) because of the force of gravity and the force of the ball on the person's hand. (The downward force of the ball on the person is the N3L reaction force to the upward push of the person on the ball.)

The vertical velocity of the ball is shown by the open circles in the graphs. In both cases, beginning at about 0.45 s (when the feet leave the ground), the vertical velocity of the basketball increases rapidly. This upward acceleration of the basketball is traceable to wrist motion, which we will call the wrist snap. Videopoint® was used to calculate the acceleration of the basketball. It showed that the upward acceleration of the basketball during the wrist snap had values up to about 70 m/s^2. That's about 7 *g*s of acceleration upward. According to N2L, this would require 8 "*g*s of force." I dislike that terminology. What is meant is that an acceleration of 7 *g*s upward requires a force of 8 *mg* (eight times the weight). One *mg* cancels the downward force of gravity and the remaining 7 *mg*s produce the acceleration. For a basketball 8 *mg*s is about 48 N (11 lbs). Later in this chapter, we have more to say about the forces involved in shooting a basketball. Finally, the open circles in figures 3.4 and 3.5 show that both Mike and I release the basketball at about 0.62 s. More importantly, the filled circles show that for both of us when the ball is released, the upward speed of the head is about zero, the head is not moving up or down. This confirms something that most basketball experts would agree on. For a jump shot, the ball should be released at the top of the jump.

There is a difference between what Mike Heary and I do with the basketball prior to the wrist snap. Figure 3.4 shows that Mike starts his shot ($t = 0$) by pushing the ball vertically upward. The ball accelerates upward rap-

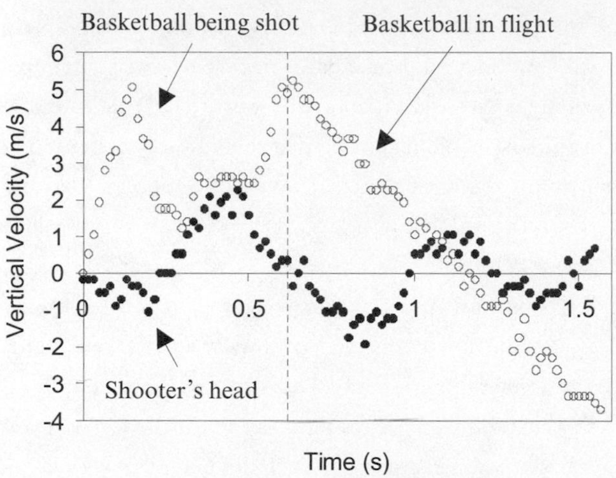

Figure 3.4. Analysis of the jump shot of a good shooter. (Picture) Open circles, position of the ball at 1/60 second intervals. (Graph) Open circles, vertical velocity of the basketball; closed circles, vertical velocity of the shooter's head

idly, then slows. (The vertical velocity goes through a maximum—the peak shown by the open circles.) The ball then travels upward with the same velocity as the head until the wrist snap begins. For my shot (fig. 3.5), the ball initially travels downward with about the same velocity as the head. If that continued, it probably would not represent bad technique. It seems reasonable for the ball to follow the motion of the head in the beginning.

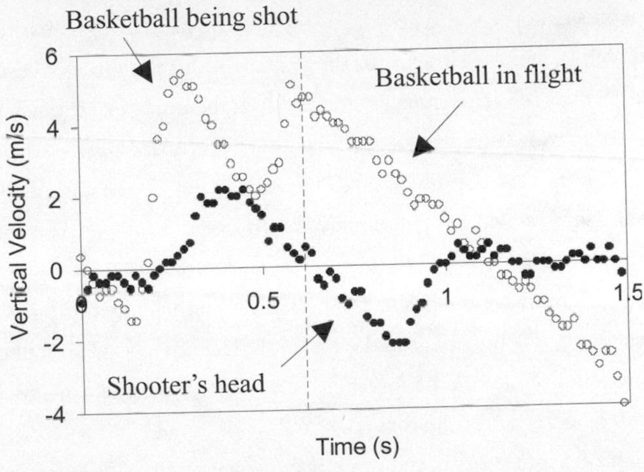

Figure 3.5. Analysis of the author's jump shot. (Picture) Open circles, position of the ball at 1/60 second intervals. (Graph) Open circles, vertical velocity of the basketball; closed circles, vertical velocity of the shooter's head

Unfortunately, beginning at about 0.1 s the ball goes downward a bit faster than does the head. Consequently, it seems that I lower the ball an excessive amount these days. That's bad technique. It takes time, and the shot starts upward from a lower point than is necessary. Both of these factors help the person on defense. They make it easier for her or him to steal the ball, or at least disrupt the shot. Perhaps this is a bad habit that I have developed since my playing days. It may be that I developed the bad habit because starting with the ball lower makes it easier to jump. Jumping is difficult at best these days because of old legs and bad knees. One good feature of my shot is that once the basketball begins to move upward, it moves vertically and backward a bit. That makes it more difficult for the person on defense since the path of the ball maximizes the distance from her or him. Getting back to the graph in figure 3.5, we see that the ball accelerates upward rapidly beginning at about 0.2 s. The velocity of the basketball then goes through a maximum similar to that for Mike Heary's. The difference between the shots is that the peak occurs at a much later time for my shot. For my shot, after the upward motion of the basketball begins, it does not travel with the head (or body) for an extended period of time. The wrist snap begins as soon as the upward speeds of the head and the basketball match. While the overall time for the shots is about the same, it could be argued that my release is quicker. For my shot, the time between when the ball begins to move rapidly upward (beginning of the first peak of the open circles) until the release point (dashed line) is shorter. The total time for my shot is the same as Mike's, so, at first sight, this would appear to be a moot point. Delaying the upward motion of the basketball, however, gives the shooter options. Since the basketball remains with the body during the initial knee bend, it is possible to fake the shot. The knee bend could be followed with a dribble or a pass rather than a shot. These options make it harder to defend against the move.

Analyzing foul shots in the same way showed that Mike Heary's foul shot is almost the same as his jump shot. He doesn't jump when he shoots foul shots but just extends his body as far as he can then releases the ball. Also, both raising the ball to the head and preparing for the wrist snap are slower. I use a foul-shooting technique that is significantly different from my jump shot. This technique should be of interest to some players because it doubles as a one-hand set shot.[7] When it is used as a set shot, the player usually jumps a little, though the hand and wrist motion remain essentially the same. Because it is different and of more general interest, my foul shot was analyzed. The resulting picture and graph are shown in figure 3.6. An additional feature appears in the graph. Besides the usual vertical velocity of the basketball (open circles) and the head (filled circles), the horizontal velocity of the basketball is plotted (open triangles).

The picture in figure 3.6 shows that this type of foul shot uses a more direct approach to the basket. The ball starts higher and travels a shorter distance before the release. According to the graph, the basketball stays at rest until about 0.15 s. During the first 0.15 s the head (filled circles) and the body move downward a small amount. The basketball then accelerates rapidly upward and to the left until a time of 0.2 s. The upward velocity then increases slowly until about 0.35 s. Between 0.2 s and 0.35 s the hand moves backward in preparation for the wrist snap. This is reflected in the negative horizontal velocity of the basketball from 0.2 s to 0.35 s. Also, during this time the head begins to move upward. At about 0.35 s the wrist snap begins. In addition to the usual vertical acceleration of the ball during the wrist snap, figure 3.6 shows that there is horizontal acceleration of the basketball (open triangles) during the wrist snap. Finally, the basketball is released at about 0.45 s.

The data (filled circles) show that the head does not stop moving upward until slightly more than 0.5 s. Consequently, the head is still moving upward when the basketball is released. This is different from a jump shot, where the basketball is released when the vertical velocity of the head is zero, at the top of the jump. This is one reason that the time required for a foul shot (or a set shot) is shorter than for a jump shot. The data show that the total time required for a foul shot or set shot (0.45 s) is about three fourths that for a jump shot (0.62 s). Further, we have seen that shooting a basketball from a lower point has both a wider range of allowed angles of approach and a larger window at all angles of approach. Consequently in cases where the height of the point of release is not important, as is the case for foul shots and many shots from three-point range, there are advantages to using a set shot.

One advantage to using a foul shot that is similar to a jump shot is that it is only necessary to develop one type of shot. The motion and release for Mike Heary's foul shot are almost identical with those for his jump shot. Consequently, he only had to learn one type of control with the hands and wrists, though the motion of the rest of his body is slightly different for his foul shot and jump shot. Another advantage of Mike Heary's foul shot is that close to the time of release (launch), there is less blocking of vision by the hands and the ball since the basketball is shot with a release point above the head.

The need to only develop one type of shot represents the strongest argument against using my foul shot/set shot technique. In most cases, however, there is a very good opportunity for a player to develop and use a foul-shooting technique like mine. The reason is that a set shot is the easiest shot to shoot. I was very young when I started playing the game. I knew the varsity plays for Wampum High School by the time I was six years old and I played in a "pee-wee" game at half-time of the varsity game when I was seven. A set shot is the only reasonable shot that a very young person can manage, because a set shot requires less use of the hands. In the case of the set shot it's possible and desirable to use the legs to give the upper body and arms some velocity. We have already seen that the head and presumably the body are not at rest when the basketball is released. Since the basketball already has the velocity of the arms, less velocity needs to be provided by the wrist snap.

In my opinion, the best way to train very young players is to teach them a set shot. It is difficult, if not impossible for children to develop a proper jump shot. I remember that the earliest that I was able to begin shooting a jump shot was about age thirteen. Developing a jump shot later is fine and highly desirable. It is easy for young children to develop a proper set shot and that can easily transition to the player's foul shot. Developing a foul shot based on a set shot rather than a jump shot is a reasonable path to take. For me there were reasons other than the fact that jump shot clones take more effort. I found it more difficult to control the basketball when it was released from above my head. Also, using a different, easier technique seemed to make it possible to relax a little more at the charity stripe.

When I started playing, using a foul shot similar to a set shot was revolutionary for a different reason. At that time, most of the players shot underhand foul shots. Rick Barry was one of the last players to shoot underhand foul shots. Developing a third shooting technique didn't make sense to me so Coach Hennon let me shoot one-hand foul shots. It has always worked well. I remember making three hundred in a row once while practicing at

Westminster College. I was the NAIA Free Throw Shooting Champion for the 1964–1965 season with a 92.2%. Finally, I'm a little surprised that the advent of the three-point line hasn't renewed interest in the set shot in the men's game. It takes less effort and is more accurate and quicker than a jump shot. In the women's game the set shot is more widely used.

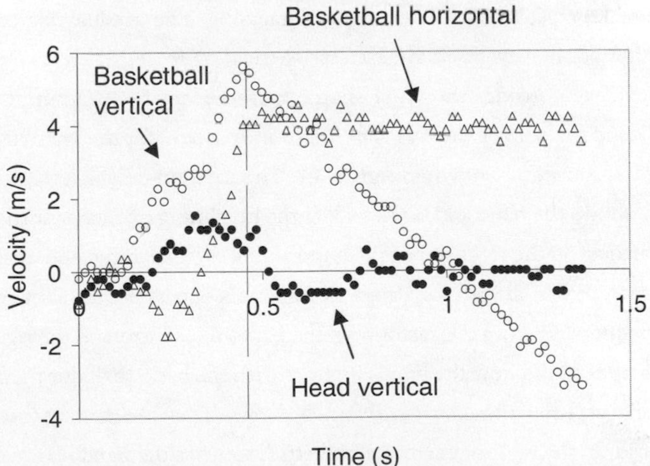

Figure 3.6. Analysis of a foul shot similar to a set shot. (Picture) Open circles, position of the ball at 1/60 second intervals. (Graph) Open circles, vertical velocity of the basketball; open triangles, horizontal velocity of the basketball; closed circles, vertical velocity of the shooter's head

There's an interesting way to locate the center of the foul line, at least for wooden basketball courts. Most wooden courts have a small hole in the floor at the center of the foul line. That is because of the way that the circle around the foul line is established. A nail is usually pushed or hammered into the floor at the middle of the foul line and string is attached to the nail. When the string is stretched, its other end is a constant distance from the nail. A pencil at the end of the stretched string is then used to trace out the circle. When shooting a foul shot, I always place the front of my right shoe at the small hole in the floor. This enables me to always shoot free throws from the same position relative to the basket and this improves consistency.

We have already considered the flight of the basketball in some detail, but the flight of the basketball is considered from a different point of view to the right of the dashed line in figure 3.6. The vertical velocity (open circles) of the basketball decreases at an approximately uniform rate, that is, there is an approximately constant downward acceleration. That is expected, since gravity is the dominant force and is approximately constant in a downward direction. The horizontal velocity (open triangles) is relatively constant because neither gravity nor the buoyant force pushes or pulls horizontally. However, careful study of the horizontal velocity in figure 3.6 reveals that it is gradually decreasing. The gradual decrease in the horizontal velocity is attributable to the drag force.

Next, we consider the wrist snap in some detail in an attempt to understand the role of the hands in the shot. A model for the wrist snap of a good shooter is shown in figure 3.7. The cartoon on the left in figure 3.7 shows the hand and basketball at the beginning of the wrist snap. The cartoon on the right shows the hand and basketball at the end of the wrist snap, that is, the release. Proper technique is one where the shot progresses smoothly from the cartoon on the left to the cartoon on the right.

To understand how the hands interact with the basketball during the wrist snap, we need to consider two new forces. They are contact forces that occur in the region where the basketball touches the hand. They are usually referred to as the normal force of the hand on the basketball, $\mathbf{N}_{\text{hand on basketball}}$, and the friction force of the hand on the basketball, $\mathbf{F}_{\text{friction}}$.

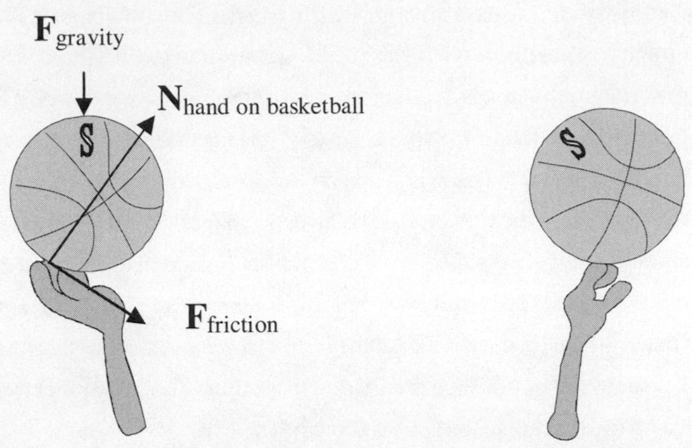

Figure 3.7. Model of the wrist snap. (Left) Position of the hand and ball and forces of the hand on the ball at the beginning of the shot. (Right) Position of the hand and ball at release.

Those contact forces are shown in the cartoon on the left in figure 3.7. The lengths of the arrows are not drawn to scale, but they do show that the total force is up and to the right.

The normal force is not normal in the sense of the word ordinary but is normal in the mathematical sense meaning perpendicular. The idea is that the normal force is a force that is perpendicular to the surfaces in contact. In this case the surfaces in contact are the surfaces of the basketball and the hand. Since the normal force is perpendicular to the surface of the basketball, the direction of the normal force on the basketball is from the point of contact to the center of the basketball (along a diameter of the basketball) as shown in figure 3.7. The normal force will be discussed further in the following chapters.

In addition to pushing in a direction perpendicular to the surface of the basketball, the hand pushes (or pulls) parallel to the surface. The force along the surface of the basketball is a special kind of friction force. If the hand does not slip on the basketball, it is called a static friction force. Static friction is a very important kind of force, especially in the game of basketball. For example, static friction is the force that is responsible for running, cutting, stopping, etc. as long as there is no slipping. In the case of shooting, it is the force that imparts rotation to the basketball. Again, it is static friction so long as the fingers do not slip along the surface of the basketball. Static friction acts parallel to the surfaces in contact when two objects—the hand and the basketball—are trying to move relative to one another but are not being successful, that is, when objects "stick" together. A more detailed discussion of the static friction force is given in chapter 5. There is no need to elaborate here because the direction of the static friction force in the case of shooting is clear. The hand is waving toward the right in figure 3.7. If the hand does not slip on the basketball the friction force must, in part, be pushing the ball to the right. Since the friction force must also be parallel to the surface of the basketball, it is in the direction shown.

The friction force, $\mathbf{F}_{\text{friction}}$, has two consequences. First, it affects the linear acceleration of the basketball. The friction force drawn in figure 3.7 would cause the basketball to accelerate to the right and down. The normal force, $\mathbf{N}_{\text{hand on basketball}}$, also affects the acceleration of the basketball. The normal force drawn in figure 3.7 would cause the basketball to accelerate to the right and up. Together, these forces act during the wrist snap to determine the launch velocity of the basketball. The friction force has a second function. It is solely responsible for the spin of the basketball. Because the friction force acts along the surface of the basketball in the direction shown in figure 3.7, it imparts counterclockwise rotation or backspin to the basketball. The cartoon on the right in figure 3.7 shows the position of the hand at release of the basketball. The rotation of the basketball can be seen by comparing the positions of the S in the cartoons on the left and right in figure 3.7.

This model explains much of what a good shooter does. For example, it shows why it is natural to impart a spin of about two revolutions per second to the ball. It is apparent from the S on the ball in figure 3.7 that the ball rotates about one eighth of a revolution during the wrist snap. Further, as we saw in figures 3.3 to 3.5, the time for the wrist snap is about 0.1 s. If a constant angular acceleration is assumed for the increasing speed of the rotation of the basketball, the final rotational speed is calculated to be 2.5 revolutions per second. If the basketball actually rotated through one tenth of a revolution in 0.1 s, the final spin would be exactly two revolutions per second. Toward the end of chapter 2 it was pointed out that it is natural to put more spin on the ball when the ball is launched with a greater linear speed. This model shows the reason. A basketball launched with a greater linear speed would remain in contact with the hand a shorter time. A shorter time would result in a greater angular acceleration and ultimately a greater rotational speed. This model also predicts a proper follow-through. The shot progresses smoothly from the cartoon on the left to the cartoon on the right in figure 3.7. This requires that the hand be moving to the right when the basketball is released. The laws of physics require that the hand continue to the right for at least a short time after the basketball is released, hence the proper follow-through.

Figure 3.3c suggests that Peja Stojakovic shoots a jump shot with a correct wrist snap and proper follow-through. The position of his right hand shows that it must have been moving to the right when the basketball was released. Because the hand is connected to the wrist, it followed an approximately circular path after the release and stopped in the position shown. A picture showing evidence of a proper follow-through for a foul shot is shown in figure 3.8. Okay, my picture does not belong in the book along with MJ and the others. A lame excuse for putting the picture in the book is that it's the only picture that I could find showing evidence of a proper follow-through after a foul shot. (I didn't look very carefully.) The real excuse is that the situation associated with figure 3.8 is important. The picture was taken just after the release of the second of two foul shots with a few seconds left in a game against Bucknell. The score is visible on the scoreboard at the top of the picture. The first foul shot tied the game and the second foul shot won the game 57–56. Michael Jordan's shot in the UNC versus Georgetown game put him on the map, and this foul shot put me on the team. I was only a sophomore at the time. Words to the wise basketball player are: Never underestimate the importance of foul shooting.

I have seen players shoot so that the basketball has no spin. Those players shot-put the basketball or shoot a knuckleball. When I see that, I immediately volunteer to play them a shooting game or two because I know that

Figure 3.8. The author shooting the second of two foul shots with a few seconds remaining on the clock in the Westminster versus Bucknell game on December 19, 1964. The first foul shot tied the game and the second won the game 57–56.

their shots will be inconsistent. One reason for the inconsistency is that the wrist is locked. The shot is caused by the normal force only. This leads to control problems because the forearm and locked wrist represent a large, relatively rigid system that must be used to accelerate the basketball. Next, though I don't have any data that verify the existence of the effect, I think that I have seen a basketball shot without spin exhibit small, random changes in trajectory characteristic of a knuckleball. Those random changes in trajectory would be attributable to air forces arising from turbulence in the air. Presumably, these changes are associated with the irregularities on the surface of a basketball in a manner similar to those associated with the stitches on a baseball.[8] While the existence of forces due to turbulence is very useful for a knuckleball baseball pitcher, it is highly undesirable for a basketball shooter where reproducibility of the trajectory is of ultimate importance. In addition, there are those who sometimes use spin and sometimes don't. This is the same as turning a force on and off because the Magnus force only acts on the ball when it is spinning. As we have seen, the trajectories are slightly different with and without spin. The different trajectories add to the variability of a shot. Finally, once an axis of rotation is established via spin, that axis will tend to remain in that orientation. This is the same phenomenon that fixes the orientation of a spinning gyro in a ship or satellite. A fixed orientation in space for a basketball in flight should also help improve the reproducibility of the shot. One only needs to watch a game or two to see that improper spin on shots is a problem that needs correcting for many of today's players. That became painfully obvious to me during the Navy versus Lehigh University game on January 15, 2005. I watched a player shot-put and miss two foul shots with the score tied at the end of the game. Navy went on to lose.

Earlier in this chapter, the importance of the release point was discussed. The release point is the position of the basketball in the cartoon on the right in figure 3.7. As was also discussed earlier, the launch of the basketball is determined by the forces on the basketball between the beginning of the wrist snap and the release. Consequently, the release point is determined once the starting position for the wrist snap is established. My favorite procedures for establishing the proper starting position for the wrist snap are as follows. For a jump shot, balance the basketball just above your forehead using the fingertips. The correct starting position for the wrist snap will be slightly in front of that position. Figure 3.3a shows an excellent starting position for the wrist snap associated with a jump shot. For a set shot (or my foul shot), lie on your back and use your fingertips to balance the basketball in front of your face for the optimum

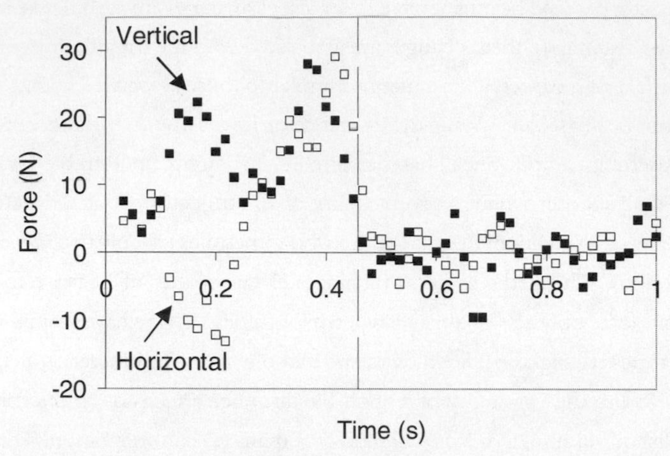

Figure 3.9. Total force of the hands on the basketball versus time for the foul shot shown in figure 3.6

position of the ball on your hand. These positions of the basketball relative to the body will result in an excellent release point with the use of a smooth wrist snap.

We end the discussion of the wrist snap with further details of the forces involved during the shot. Videopoint® was used to calculate the acceleration associated with the foul shot shown in figure 3.6. The acceleration was multiplied by the mass of the basketball to obtain the total force on the basketball. The weight of the basketball was subtracted from the total force to eliminate the effect of the force of gravity. This gave the total force of the hand on the basketball and includes both the normal and friction forces. The vertical (filled squares) and horizontal (open squares) portions of the total force of the hand on the basketball are plotted in figure 3.9.

The data to the right of the vertical dashed line at about 0.45 s are centered on zero. That is expected since the hand is not in contact with the basketball beyond the release point.

The data to the left of the vertical dashed line, when the basketball is in contact with the hand, have several interesting features. The vertical component of the force (filled squares) shows a double peak. The first peak (at about 0.15 s) causes the initial rapid rise of the vertical velocity shown in figure 3.6.

The maximum value of the force of the hand on the basketball during this phase of the shot is about 22 N (5 lbs). After that the vertical force decreases. This causes the slowly increasing vertical velocity between about 0.2 and 0.35 s in figure 3.6. The second peak in figure 3.9 is associated with the wrist snap. The maximum vertical force during the wrist snap is about 28 N (6 lbs). This is the cause of the rapid upward acceleration of the basketball during the wrist snap. The horizontal component of the force of the hand on the basketball (open squares) beginning at about 0.1 s is negative. This negative component causes the increasingly negative horizontal velocity of the basketball, pulling the basketball away from the defender in the case of a set shot. The horizontal component of the force then becomes positive and causes the horizontal acceleration of the basketball during the wrist snap. The maximum horizontal force during the wrist snap is about 30 N. The horizontal and vertical forces exert a maximum total force on the basketball of about 42 N (9.4 lbs). On the one hand (no pun intended), this is a fairly large force. It shows why young players cannot get the basketball to the hoop from as far away as the foul line. It also shows why strength is important to a shooter. On the other hand, for players like Wilt or Shaq, it is a very small force compared with what they are capable of exerting. It requires careful training for them to develop the ability to control the basketball with such small forces.

We have been analyzing a basketball shot in reverse order. We started with the flight of the ball, then considered the role of the hands and arms in the launch of the ball. We continue that sequence by next considering the role of the rest of the body in shooting. The role of the body should not be underestimated. This may sound strange, but I think that the rest of the body is more important than the hands and arms in shooting. The reason for that statement is that during high school I first noticed that I had more trouble shooting if my legs or ankles were hurt than if I had a sore finger or other small problems with my hands. I suppose that it's debatable whether the body or the hands and arms are more important in shooting. However, what's certain is that the rest of the body is important, so let's analyze its role.

We start by considering jump shots and the horizontal motion of the rest of the body. When practicing to improve reproducibility of my jump shot, I always tried to jump straight up. Figure 3.3a shows that that is what Rick Barry did. The idea is to try to eliminate any horizontal velocity of the body. Horizontal drift is difficult to control and usually ends up being random. That gives a different perspective to each shot and tends to decrease accuracy. To increase accuracy, a shooter should jump straight up whenever possible no matter how fast she or he is moving horizontally before starting the shot. The videos showed that Mike Heary does that from all distances from the hoop except in the three-point range where he moves forward a few inches during the shot. Until I watched some professional players closely, I was bothered by the amount of forward drift that I have these days even for jump shots from the foul line. I have noticed, however, that a lot of professional players move forward a large distance during a jump shot, at least for shots from beyond the foul line. The reason for the extra forward drift is that we're using the legs to help us give the basketball some of its velocity. This was discussed earlier in the chapter in the context of set shots. Since the basketball already has our forward velocity, less forward velocity needs to be given to the ball via the wrist snap. Moving the body forward during a shot has a flaw. It helps the person playing defense because it decreases the distance between the shooter and the defender. In the case of extreme forward movement, the shooter can commit a charging foul.

There is another reason for trying to go straight up on a jump shot. It helps establish a consistent release point. As discussed earlier in this chapter, after observing a few shots, the person on defense identifies the standard release point. One way to defeat the defender has already been discussed; that is, to raise the basketball a little to achieve a higher release point. Another way to defeat the person on defense is to jump differently. The classic different jump shot is the fade away. In that case, the shooter jumps backward a little. It's also reasonable to jump to the left or right a little. Michael Jordan was a master of the fade away or variable jump. Careful study of figure 3.3b reveals that he is not perfectly vertical for the shot. In fact, he is more vertical in that picture than in any of the many other pictures that I looked at. On the basis of those pictures I was forced to conclude that his jumps are seldom the same. The variability of his jumps adds significantly to the difficulty of stopping him.

It is worth developing these techniques to defeat the defender. Causing a perplexed or frustrated look on the face of the defender adds a lot of fun to the game. The problem is how to develop the ability to make adjustments. In my opinion, the best training is very little training or no training at all. The majority of shot development should be done by trying to establish a standard shot where there is no horizontal drift of the body. This is consistent with the fundamental practice principle that a player should always practice shooting like she or he plays, that is, the player should always practice under simulated game conditions. A majority of the shots in a game are taken where the player can jump straight up or even a bit forward, without the need for small up, back, left, or right adjustments. This implies that the majority of shots in practice should be standard shots. The fraction of adjusted shots should be roughly proportional to the fraction of adjusted shots in a game, which is usually a small number. In a game, there is little time to think or make conscious adjustment. A game is mostly played via reflex. It is during practice that time is available to develop the skills and reflex necessary to play the game. Consequently, the fundamental practice principle needs to be modified. A player should practice like he plays, but he needs to think about the process at least part of the time. That includes a continuous monitoring of not only whether the launch velocity and spin are correct (whether the ball goes in softly), but also of the release point and release time, and whether the body motion is correct. Thinking is the key. If two players have the same physical skills, the one that thinks will always win.

There is another way to increase the height of the release point in an attempt to defeat the defender. Both Rick Barry and Michael Jordan did it. To see what they did, look at their feet in figures 3.3a and 3.3b. In both cases the toes are pointed downward. It turns out that maximum extension of the legs and feet downward give maximum upward height to the hands. One way to look at it is that they don't have to lift all of the mass the full height during the jump. The toes and feet remain lower. Another way to look at it is that they are making the center of mass of their bodies as low as possible. The center of mass is where all of the mass of the body appears to be concentrated from a physics (force) point of view. When a player jumps, the center of mass accelerates as though all of the forces are acting on all of the mass at that point. Once the player leaves the ground, the center of mass of the player follows a trajectory determined primarily by gravity. For this discussion, we will assume that the air forces are negligible.

The position of the center of mass of the whole body is determined by where the masses of the parts of the body are located. For example, for equal masses, the center of mass is midway between them. For unequal masses, the center of mass is closer to the heavier mass. Suppose that equal masses are on a massless vertical stick that has been thrown straight up and is at its highest point. The top mass is glued to the stick and the bottom mass is movable. Next, suppose that when the vertical stick is at the highest point the bottom mass is pushed down by the stick. The center of mass moves down along the stick half the distance that the bottom mass moves down. However, the position of the center of mass cannot change position in space. Consequently, the stick and hence the upper mass must move upward exactly the right amount (half the distance that the bottom mass moves down) to keep the center of mass in the same position in space. If the upper mass represents the hands of a shooter and the lower mass represents the feet, this shows why it is important to extend the legs and feet downward as much as possible. Extending the legs and feet downward enables the player's hands to reach the highest possible point. One of the reasons for this discussion came to my attention while I was helping Coach Ridl at a basketball camp for a few days one summer. There was a player there who insisted on pulling his legs upward during a jump shot. I think that his idea was that he was getting extra "kick" propelling him upward. I was not able to convince him that that was bad physics and that what he was really doing by pulling the legs upward was to move the center of mass upward and hence his hands downward.

All of the shots that we have analyzed thus far in the book get nothing but net. In the next chapter, we'll consider some shots where the basketball goes in via bounces from the rim or the backboard.

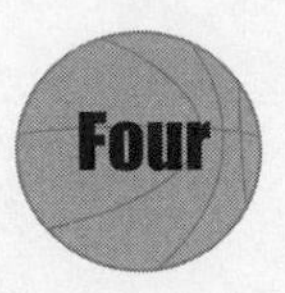

Basket Case

Most basketballs that go through the hoop don't get nothing but net. The ball usually bounces off the rim[1] or the backboard before going in. There is one shot like that that has made the TV highlights since March 2005. It is Patrick Sparks' shot at the end of the Michigan State-Kentucky elite eight game of the 2005 NCAA Men's Tournament. Sparks played for Kentucky and sent the game into overtime with about 1 second left by means of a three-point shot from the top of the key. The ball went in after bouncing off both the hoop and the glass. The shot added drama to the game and gave the Wildcats hope but they went on to lose 94–88. We will say more about that shot later in this chapter. To set the stage, we first consider some of the ways that a basketball can make its way through the hoop via the bounce. We start by ignoring the backboard and allowing the basketball to hit the rim. The situation is complicated because a basketball can hit the rim in many ways. It can hit at an infinite number of different points with an infinite number of different velocities, and that's just for straight-on shots. However, with a definition and a few restrictions, the bounces can be categorized.

We first locate the point of contact of the ball with the rim by yet another angle, the angle to the point of contact, θ_{PC}. As shown in figure 4.1, θ_{PC} is the angle from a horizontal line to the left of the center of the rim to a line from the center of the rim to the point of contact.

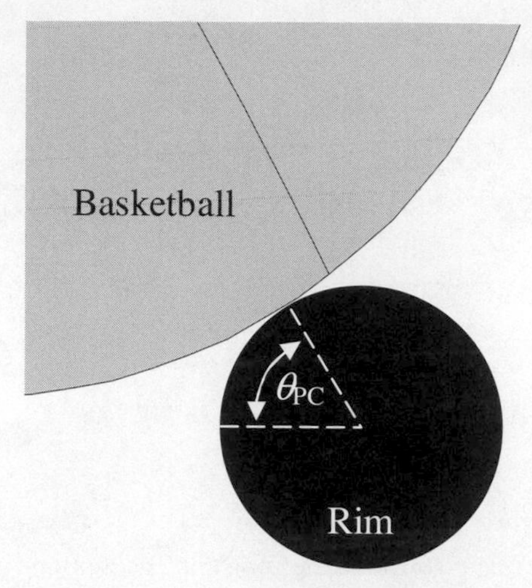

Figure 4.1. Definition of the angle to the point of contact, θ_{PC}—A basketball and a cross section of the rim are shown. The rim is drawn larger than normal to show θ_{PC}. The front of the rim is the leftmost point on the rim since, for most of the discussion, the basketball travels from left to right.

We only consider basketballs that hit the top half of the rim, that is, basketballs with values of θ_{PC} between 0° and 180°. Only considering those collisions is reasonable since we know what happens when the ball hits the underside of the rim. That kind of shot can be embarrassing, in particular, if you're all alone on the front end of a fast break. The only thing more embarrassing is being all alone on a fast break, making the shot, then realizing that it was the wrong basket.

In the first part of this chapter, we only consider basketballs that hit the rim with a velocity between horizontal to the right and vertically downward. Thus, we ignore shots where the ball has an upward component of velocity, so we are restricting ourselves to outside shots. We also ignore shots that approach the rim from the right. Shots that approach the rim from the right become important when basketballs bounce off the backboard or off the back of the rim toward the front of the hoop. We consider that later in the chapter. Finally, we assume that the bounce is perfect. The ball bounces away from the rim in the same way that it approaches the rim, except that it bounces away on the other side of the perpendicular to the surfaces in contact. In the language of physics, we assume that the angle of incidence equals the angle of reflection. This is only correct if the rim is frictionless and the basketball is perfect. Even though the rim is frictionless, it can exert a normal force on the basketball since the normal force is perpendicular to the surfaces in contact. Consequently, the assumption is that the bounce is caused only by the normal force. As discussed in the next chapter, a perfect basketball is one where there are no internal friction forces. A perfect basketball dropped vertically onto the floor would bounce back to the same height that it is dropped from. (This would only occur if there were no atmosphere.) Another consequence of ignoring all kinds of friction is that the path of the bounce from the rim is reversible.

With these assumptions, the infinite number of rim bounces can be grouped as shown in figures 4.2 to 4.4. (Note: If you read this chapter carefully, you will probably get as much of a headache as I did when I wrote it. However, there should be some gain from the pain.) In each of the figures, the solid dark circle represents the rim. The rim and basketball are drawn out of proportion to help clarify the point of contact between the ball and the rim. In each of the figures two extreme incident velocities are shown. One of the velocities is shown via a solid arrow, labeled V, pointing downward. Velocity V represents a basketball that approaches the rim vertically from above. In figures 4.2 and 4.3 a horizontal dashed arrow, labeled H, points to the right. Velocity H represents a basketball that is moving horizontally to the right when it hits the rim. The dashed arrows in figure 4.4 are not horizontal because it is impossible for a basketball traveling to the right to hit the back side of the rim. The reason is that the ball cannot go through the steel of the rim to hit the backside of the rim where the point of contact is located. In figure 4.4, the leftmost dashed arrows are labeled MH and represent the most horizontal direction of travel that a basketball can have if it is to hit the rim at the point shown. Mathematical details of the angles associated with the incident velocities are given in appendix V.

The physics also places a restriction on how a basketball can approach the rim at the back of the hoop. A basketball traveling toward the basket relatively horizontally cannot hit the front of the rim at the back of the hoop because the front of the hoop is in the way. For straight-line travel (infinite speed), a basketball cannot be traveling less than 22.2° below the horizontal and hit the rim at the back of the hoop at $\theta_{PC} = 0°$. The minimum angle of travel gradually decreases to zero as θ_{PC} increases to 90°.

$\theta_{PC} = 0°$

We now consider what happens when a basketball hits the front of the rim as shown in figure 4.2a. This point of contact is described by $\theta_{PC} = 0°$. In this case, the center of the basketball is in the plane of the hoop when the basketball hits the rim. A basketball that approaches the rim horizontally (H) bounces straight back. That bounce is represented by the dashed line with an arrowhead at each end. The solid arrows show that a basketball that approaches the rim from directly above (V) continues on its downward path since it just grazes the rim. All basketballs with in-between directions of approach bounce downward and to the left. Two typical in-between bounces are represented by the dash-dot (1) and the dash-double-dot (2) arrows. The bad news is that if a basketball hits the front of the rim *at the front of the hoop*, it always misses. The good news is that if a basketball hits the front of the rim *at the back of the hoop*, it always goes in. Although it's premature to generalize, these results suggest that a first rule of shooting might be "Don't shoot short."

Allowing a basketball to hit the front of the rim at the back of the hoop greatly increases the probability that the ball will go in versus when it is required to get nothing but net. I made some calculations for my free throws (shots released from a horizontal distance from the backboard of 4.2 m [14 ft] and a height of 2.4 m [8 ft] above the floor), allowing the basketball to go in by way of a bounce off the back of the rim at $\theta_{PC} = 0°$. I found that the minimum allowed angle of approach was 19.0° for men and 18.0° for women. These values are much lower than the corresponding values of 28.8° and 27.8° quoted in chapter 3 for foul shots that get

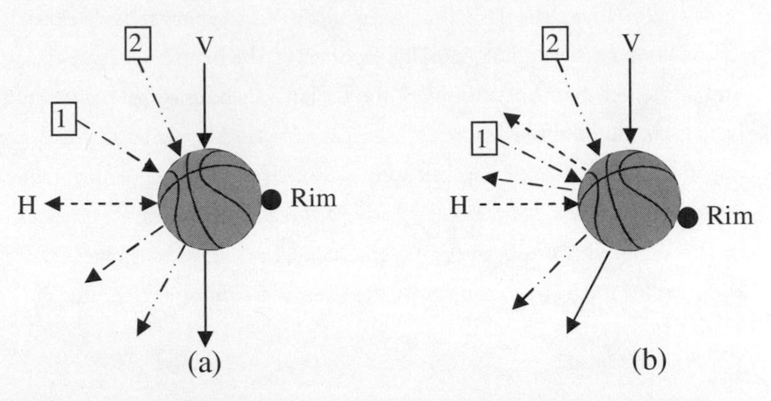

Figure 4.2. Basketball colliding with a rim with an angle to the point of contact of 0° (a), and between 0° and 45° (b)

nothing but net. The extra (almost 10°) allowed angles of approach make it much easier to make a shot when the ball is allowed to hit the rim.

θ_{PC} *between 0° and 45°*

In figure 4.2b, the basketballs hit the rim with a θ_{PC} of about 15°. It is typical of what happens when θ_{PC} is between 0° and 45°. The dash-dot arrows show that a basketball that approaches the rim relatively horizontally (bounce 1) bounces upward and to the left. The dash-double-dot arrows show that a basketball that approaches the rim relatively vertically (bounce 2) bounces downward and to the left. Because of the reversibility of the bounces, the path shown by bounce 1 represents two possible bounces. Any basketball that hits the rim at the front of the hoop with θ_{PC} between 0° and 45° misses. All bounce away from the basket. This is more evidence that a player should try to not shoot short.

The situation is more complicated if one of the basketballs in figure 4.2b hits the rim at the back of the hoop. If the ball approaches the hoop relatively vertically (bounce 2), the ball bounces downward and to the left. That represents more than half of the bounces because relatively horizontal approaches do not occur because the front of the hoop is in the way.

For those bounces, the basketball goes in. If the ball approaches the back of the hoop relatively horizontally (bounce 1), the ball bounces upward and to the left. For those bounces, the basketballs bounce back over the basket. Some of those bounces miss because they proceed to the left beyond the front of the hoop, but some go in. Overall, there is a much better than 50-50 chance of the ball going in if the ball hits the rim at the back of the hoop with θ_{PC} between 0° and 45°. This begins to suggest that shooting for the front of the rim at the back of the hoop is desirable.

θ_{PC} *from 45° to 90°*

Bounces for $\theta_{PC} = 45°$ to $\theta_{PC} = 90°$ are represented by figure 4.3. By comparison with figure 4.2, there are two new features. First, no basketballs bounce downward. Whether they contact the rim at the front or back of the hoop, none of the basketballs automatically go in. Second, half of the bounces have a velocity component to the right (dash-double-dot arrows in figure 4.3b) and half have a component to the left (dash-single-dot arrows in figure 4.3b).

If one of these basketballs hits the rim at the front of the hoop, the half with a velocity component to the left do not go in. The other half has a chance of going in since the ball proceeds over the hoop after the bounce. This is the first example of a short shot that has the possibility of going in. For reasons that are given at the beginning of the next chapter, my son and I watched a Wizards-Mavericks game closely during the 2005 season. I remember a shot that hit the front of the rim, bounced almost straight up then went through the hoop. That was an example of bounce 2 in figure 4.3b.

The odds that a basketball with θ_{PC} between 45° and $\theta_{PC} = 90°$ will go in are better if the shot is taken from the vicinity of the key. The vicinity of the key is defined as being in front of the basket or a few degrees (yet another angle) to either side, that is, the middle of the court. Basketballs shot from the vicinity of the key that would have gone beyond the back of the hoop usually bounce off the backboard/glass. The backboard reflects the ball back over the hoop so that there is another chance that it will go in.

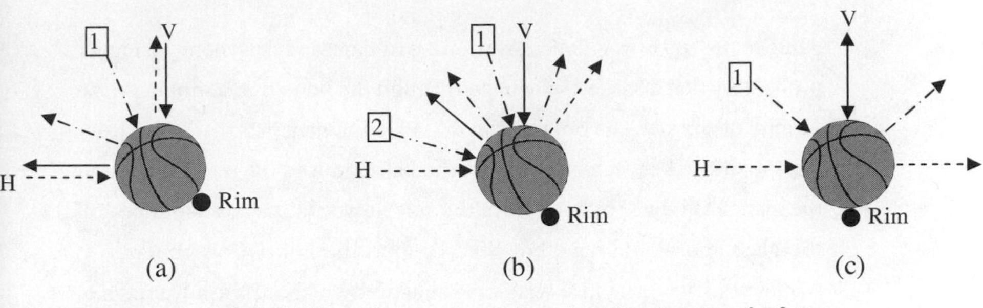

Figure 4.3. Basketball colliding with a rim at an angle to the point of contact of 45° (a), between 45° and 90°(b), and 90°(c)

If a basketball hits the rim at the back of the hoop with $\theta_{PC} = 45°$ to 90°, there is a reasonable chance that it will go in. The bounces with a velocity component to the left after the bounce have a chance of going in because the basketballs bounce back over the hoop. The bounces with a velocity component to the right miss if the shot is from the side. The number of those bounces is smaller than the number of equivalent bounces from the front of the hoop. This happens because they require relatively horizontal approaches that are prevented by the presence of the front of the hoop. In addition, if any of the basketballs that bounce to the right are shot from the vicinity of the key, they also have the possibility of going in because of the backboard.

All things considered, if a basketball hits the front part of the rim (θ_{PC} between 0° and $\theta_{PC} = 90°$), it is favorable to hit the back of the hoop rather than the front of the hoop. This confirms our first rule of shooting, "Don't shoot short." There is a danger in shooting long, though. We have seen basketballs that bounce off the rim to the right. If those basketballs hit high on the rim at the back of the hoop, they must miss if the shot is from the side. If the basketball is shot too long from the side, the fans for the opposing team once again get to practice their chant "Air ball, air ball . . ." This suggests a second rule of shooting: "If you shoot from the side, don't shoot long."

The corollary is that there is an advantage to shooting from the vicinity of the key because of the presence of the backboard. About an hour after writing this, I watched the Michigan State-Kentucky game that is mentioned at the beginning of this chapter. Patrick Sparks, who tied the game at the end of regulation, must have understood that there is a higher probability that a basketball will go through the hoop if it is shot from the vicinity of the key. The only reason that his shot went in is that it was from the top of the key. Before going in, the ball bounced off the rim then off the glass. Had the shot been from the side, it would not have bounced off the glass and would have bounced beyond the hoop. Consequently, it would not have gone in. I wish that I had known about the advantage of shooting from the vicinity of the key when I was playing. My last shot as a college player bounced off the front of the rim and continued to the backboard. Unfortunately, the shot was from the left wing so the basketball bounced off the backboard away from the basket. Had the shot been from the vicinity of the key the basketball might have gone in. That was an important shot. There were only a few seconds left and the winner of the game advanced to the elite eight of the 1967 NAIA Men's tournament. We lost 55–53.

θ_{PC} *from more than 90° to 180°*

Bounces for θ_{PC} greater than 90° to less 180° are represented in figure 4.4. By comparison with figures 4.2 and 4.3, two different features are apparent. The first different feature was discussed just before the section labeled $\theta_{PC} = 0°$. The velocity cannot be perfectly horizontal for a basketball to hit the rim at θ_{PC} between 90° and 180°. That is because it is impossible for the ball to go through the steel of the rim. An approaching basketball must be traveling downward and to the right. A few examples of the most horizontal (MH) that the velocity can be are shown in figure 4.4. The second different feature is that some basketballs continue downward and to the right after the bounce. This happens about half of the time that a basketball strikes the rear top quarter (θ_{PC} from 90° to 180°) of the rim.

A basketball that hits the rear top quarter of the rim at the front of the hoop goes directly into the basket half of the time. The other half of the basketballs continue upward and to the right. Those basketballs also have a chance of going in because they proceed over the hoop. Basketballs that are shot from the vicinity of the key have a further added mode of success caused by the presence of the backboard. Overall, the chance of the basketball going in if it hits the rear top quarter of the rim is much greater than 50–50. This validates my foul-shooting mantra—"right up over the rim." I didn't realize it at the time, but shooting the ball right up over the rim at the front of the hoop is what the physics says to do. This suggests that we might write a third rule of shooting: "Try to get the ball right up over the front of the hoop."

I suppose that it is debatable, but I am convinced that Christian Laettner did precisely that with the shot that won the NCAA East regional championship game for Duke over the University of Kentucky on March 28, 1992. The game was tied 93–93 at the end of regulation. In overtime Kentucky was ahead 101–100 with 19.6 s on the clock. Laettner sank two free-throws with 14.1 s remaining to put Duke ahead. Sean Woods of Kentucky countered with 12-footer off the glass with about 6 s on the clock then Duke called a time out with 2.1 s remaining. Grant Hill threw a baseball pass about 23 m (75 ft) from the baseline to the other foul line that Laettner caught cleanly with his back to the basket. He faked right, took a dribble then took a turnaround jump shot from just beyond the foul line. I have concluded from watching the video that the basketball just grazes the rim before going in. A good model for such a shot is a path close to the MH path in figure 4.4b.

If a basketball is shot from the side and hits the rear top quarter of the rim at the back of the hoop, it misses. This supports our second rule of shooting: "If you are shooting from the side, don't shoot long." However, a new wrinkle is added for basketballs shot from the vicinity of the key. Those basketballs cannot hit the rear top quarter of the rim at the back of the hoop if the basketball is shot from the vicinity of the key. They are blocked from that area of the rim by the steel plate that connects the hoop to the backboard. In that case the ball bounces generally upward and to the left because of the steel plate and the backboard. This usually puts the basketball back over the hoop and thus the basketball has a chance of going in.

Several parts of the discussion so far suggest a fourth rule of shooting: "Whenever possible, shoot from the vicinity of the key." We could

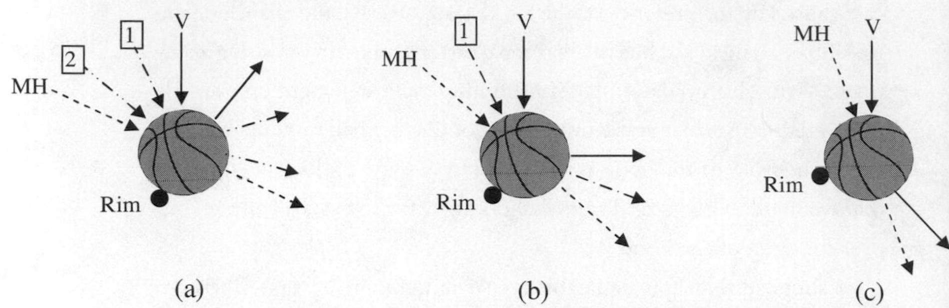

Figure 4.4. Basketball colliding with a rim with θ_{PC} between 90° and 135°(a), equal to 135° (b), and between 135° and 180° (c)

complete the set of rules by going beyond the scope of the book and write "Don't shoot to the left or right." While these rules might seem trivial—okay, they are trivial—they represent a useful guide to developing good shooting technique and game strategy.

The first rule tells a player not to shoot short. When practicing, it's important to work on developing a shot that always goes beyond the front of the rim. The second rule of shooting tells a player not to shoot long when shooting from the side. Again, that's because a straight-on long shot from either the wing or the corner has no chance of going in.

Rules one and two together imply that shooting from the wing or the corner are good tests of a shooter's ability. When shooting a straight-on shot from the wing or the corner, there is no chance that the backboard will intervene and convert a bad shot into a good one. This suggests that practicing shooting from the wing or the corner has added effectiveness in developing a good outside shot. In fact, shots from the wing are not the best test of a shooter's ability. There is always the chance that the shot will "bank" (off the glass) in. It's not that good shooters don't or shouldn't bank shots from outside. A few minutes after writing this in December 2005, I was watching a game on TV and saw Adam Morrison of Gonzaga University bank in a three-point shot from the wing with 2.5 seconds left in a game against Oklahoma State. The shot gave Gonzaga a 64–62 comeback win. The master of the bank shot was Don Hennon, my first high school coach's son, who was an All-American at the University of Pittsburgh in the late fifties. Bank shots are worth practicing and using. However, only shots from the corner require the shooter to deal solely with the hoop. As such, only shots from the corner are a true test of a shooter's ability.

The third rule is the first positive rule. It tells what to do rather than what not to do. This rule may also be the best guide for practicing shooting. To me, trying to put the basketball just over the front of the rim is the Holy Grail of shooting. The fourth (and second positive) rule suggests game strategy. The physics says that shots from the vicinity of the key should be the highest percentage outside shot. It would be interesting to learn whether statistics support this. My guess is that the effect is not very large for good shooters. However, the implications of the fourth rule are clear. If I were coaching a team composed of poor shooters, I would devise as many plays as possible to result in a shot from the vicinity of the key. No matter what the quality of the shooters on the team, I would always call for a shot from the vicinity of the key if we needed points with time running out at the end of the game. I would make that call even for a team of good shooters because the difference between winning and losing is sometimes very small. The alleged advantage for shots from the vicinity of the key also has an interesting implication for defense. If the other team has the ball with time running out, I would coach my team to try to not allow shots from the vicinity of the key—force the team on offense to the sides.

Friction and Spin

We now consider bounces where there is friction between the rim and the basketball. As discussed in chapter 3, if the basketball does not slip on the surface of the rim, the friction force, $\mathbf{F}_{\text{friction}}$, is known as static friction. For bounces, the friction force is parallel to the surfaces of the rim and basketball at the point of contact. The direction of the friction force is opposite to the direction of the velocity of the surface of the basketball at the point of contact. The basketball could slip on the surface of the rim. In that case, the friction force would be known as kinetic friction. The primary difference is that the kinetic friction force would be weaker than the static friction force.

If a basketball is not spinning, as is the case for some "bricks," the velocity of a point on the surface is the same as the velocity of the center of mass of the basketball. Consequently, when a nonspinning basketball hits the rim, the direction of the force of friction is opposite to the velocity of the basketball. An example, based on the horizontal approach in figure 4.3c, is shown in figure 4.5a.

Friction has two effects. First, friction usually slows the basketball. Figure 4.5b shows the basketball just after the collision with the rim. The slowing of the basketball is represented by the shorter dashed (velocity) arrow in figure 4.5b. One case where a basketball does not slow is a head-on collision when the velocity of the basketball can be drawn through both the center of the basketball and the rim. In that case, friction does not act and a perfect, nonspinning basketball bounces straight back with its speed unchanged.

The slowing of a basketball by friction increases the chances of its going in. For example, a basketball sometimes bounces (to the left) off the back of the hoop then travels beyond the front of the hoop. Another example is when a basketball bounces (to the right) off the front of the hoop, then travels beyond the back of the hoop. If the speed after the bounce is reduced, the ball doesn't travel as far. As a result, the probability that the ball will go in is increased. These ideas are related to the hoopothesis that good shooters minimize the speed of the basketball at the rim. The effect of friction is to decrease the speed further. This helps any shooter.

Second, friction changes the spin of the ball. If the bottom portion of the basketball contacts the rim, friction gives the basketball a clockwise

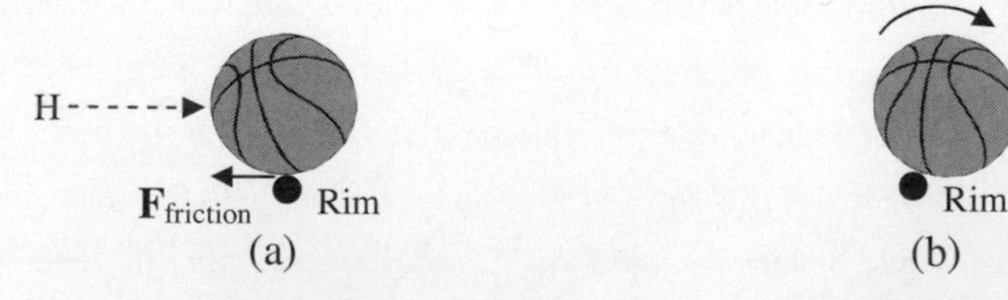

Figure 4.5. The collision of a nonspinning basketball with a rim with friction at $\theta_{PC} = 90°$; during the collision (a), and just after the collision (b). Only the force of friction is shown.

(CW) rotation. An example is shown in figure 4.5a. The clockwise rotation after the bounce is represented by the curved arrow in figure 4.5b. The basketball can be said to have topspin after it hits the rim. If the top portion of the basketball contacts the rim, friction gives the basketball a counterclockwise (CCW) rotation.

Let's consider the bounces shown in figures 4.2 to 4.4 assuming that the rim is not frictionless. The rim gives a CW rotation to the basketballs in figures 4.3c and 4.4a–c. However, the basketballs in figure 4.2a receive a CCW rotation. About half of the basketballs in figures 4.2a–b and 4.3a–b receive a CCW rotation. This seems academic, but there are a couple of reasons that it's important. First, the rotation of the basketball after the first bounce influences subsequent bounces of the ball from the rim, backboard, or steel plate. Second, it helps categorize what happens when a spinning basketball hits a rim with friction.

We now consider the bounce of a spinning basketball from a rim with friction. This is a reasonable next step because a properly shot basketball has backspin. We can determine the direction of the friction force of the rim on the basketball as follows. If the surface of the basketball moves to the right relative to the surface of the rim, the force of friction is to the left and vice versa. A tricky feature is associated with the velocity of the surface of the basketball. If the speed of the center of the basketball is slower than the speed of the surface, then one part of the surface can be moving through the air in a direction opposite to the velocity of the center of the ball.

This is represented in figures 4.6a and b. The basketball shown in figure 4.6a is only spinning. Since it is spinning CCW, the top surface is moving to

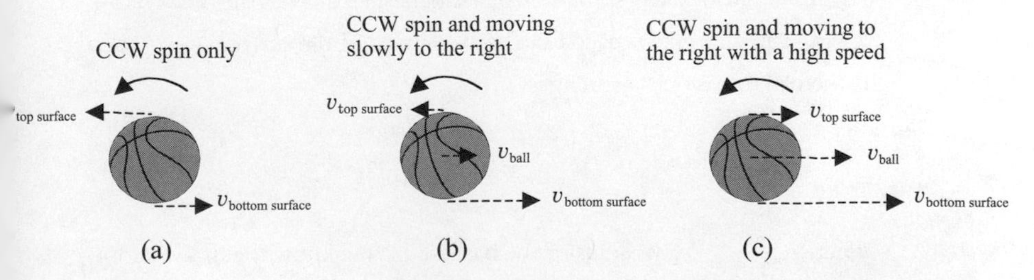

Figure 4.6. Velocities of a basketball with CCW spin only (a), CCW spin while moving slowly to the right (b), and CCW spin while moving to the right at a high speed (c)

the left and the bottom surface is moving to the right. The basketballs in figures 4.6b and c are rotating CCW and moving to the right. The basketballs in figures 4.6b and c can also be described as having backspin.

In figure 4.2b, the speed of the basketball is slower than the speed of the top surface due to rotation. Consequently, the speed of the top surface dominates and, at any instant, the top surface is moving to the left through the air even though the basketball is moving to the right. The arrow at the top of the basketball in figure 4.2b shows this. It is both to the left and smaller than the arrow at the top of the basketball in figure 4.2a. The interesting consequence is that, if the top surface collides with the rim, the basketball speeds up because the friction force is in the same direction as the velocity of the basketball. Conversely, the velocity due to rotation of the bottom surface is in the same direction as the velocity of the basketball. This makes the bottom surface move to the right faster than the speed of the basketball. The arrows at the bottom of the basketball in figures 4.2a and 4.2b show this. If the bottom surface collides with the rim, the basketball slows down.

In figure 4.2c, the speed of the basketball is faster than the speed of the top surface. Consequently, the speed of the basketball dominates and at any instant, both the top and bottom surfaces are moving to the right through the air. The speeds of the top and bottom surfaces are slower and faster than the speed of the center of the basketball. The arrows in figure 4.2c show this.

The different possibilities for the velocities of the top and the bottom of the basketball would make it complicated to analyze the collisions of the ball with the rim, but we can show that we don't have to worry about this much, at least for the first collision of a shot basketball with the rim. The following equation allows us to calculate the speed of the surface, v_{surface}, of a basketball that is only spinning

$$v_{\text{surface}} = \omega R_{\text{basketball}} \tag{4.1}$$

where $R_{\text{basketball}}$ is the radius of the basketball. We know that good shooters shoot with a backspin of about $\omega = 2$ revolutions per second.

Multipling by 2π to get the correct units for ω and using $R_{\text{basketball}} = 0.12$ m, we find that $v_{\text{surface}} = 1.5$ m/s. v_{surface} is slower than the approach speed for most shots. For example, figure 3.2 shows that the approach speed for foul shots is on the order of 5 m/s. Consequently, for most shots, the velocities through the air of the surface of a basketball are dominated by the velocity of the center of the basketball.

This implies that we can use the direction of the velocity of the center of the basketball to deduce the direction of the friction force on a basketball with backspin that collides with a rim. The direction is the same as that for a nonspinning basketball. The difference is that different parts of the basketball travel with different speeds through the air. This implies that the strength of the friction force will be different when different portions of the ball hit the rim. For the example shown in figure 4.6c, the bottom of the basketball travels faster through the air than the top of the ball. Consequently, the collision of the bottom of the ball with a rim will be more violent than a collision of the top of the ball. The result is that the static friction force on the bottom of the basketball is larger. Consequently, the basketball is slowed more if the bottom of the ball hits the rim. In addition, the rotation is affected more strongly if the bottom of the ball hits the rim. These results can be generalized. A basketball is slowed more and the rotation is affected more if the bottom *portion* of the ball hits the rim.

Which basketballs with backspin are slowed more can be determined from which nonspinning basketballs receive CW and which receive CCW spin. Nonspinning basketballs receive CW spin because the bottom portion of the basketball collides with the rim. Examples are the basketballs represented by figures 4.3c and 4.4. Those basketballs are slowed more if they have backspin before they hit the rim. This is particularly helpful to the shooter for the basketballs that bounce upward and to the right off the front of the hoop. More of those bounces don't make it to the back of the hoop and thus more basketballs go in. Nonspinning basketballs receive CCW spin because the top portion of the basketball collides with the rim. Those basketballs are slowed less if they have backspin. However, very few of those bounces help the shooter. For example, all of the bounces in figure 4.2 that receive CCW spin go in anyway so it doesn't matter whether they have more or less spin. Other bounces can be analyzed similarly. So, in general, basketballs with backspin that bounce off the rim are slowed more than if they are not spinning. Because slower speed is desirable, this is another reason that putting backspin on the basketball helps the shooter.

The second effect of backspin is to alter the way that spin is changed by the bounce. This is more directly related to which nonspinning basketballs receive CW and which receive CCW spin. A basketball moving to the right with backspin has CCW spin. If the basketball would have received CCW spin if it were nonspinning initially, it spins CCW faster after the bounce. However, the change in spin is less than if it were nonspinning. If the basketball would have received CW spin if it were nonspinning initially, friction attempts to reverse the direction of the spin. Consequently, the CCW spin is either slowed or reversed. However, the change in spin is greater than if it were nonspinning. As we see in the next chapter, ordinary bounces of these basketballs result in CW spin.

The primary effect of rotation after the bounce is to affect subsequent bounces of the basketball. That's important since shots can be made via multiple bounces from the hoop, backboard, or steel plate. We will not consider multiple bounces in any detail, in part, because any subsequent bounce from the top half of the rim can be analyzed using the techniques already discussed in this chapter. For example, suppose that the ball bounces off the back of the hoop to the front of the hoop. The new feature is that the ball approaches the rim from the right rather than from the left. It is possible to see what happens in that case by turning over the page containing figures 4.2 to 4.5 and looking at the figures through the page. There is one subtlety to watch out for. If a basketball is rotating CCW and moving to the right, it has backspin. If a basketball is rotating CCW and moving to the left, it has topspin. There is one general comment, however, that can be made about subsequent bounces from the rim. Because of friction, most bounces slow the basketball further. It follows that each subsequent bounce back over the hoop increases the probability that the ball will go in.

Before going on, we consider an example of a subsequent bounce. Suppose that a basketball with CW spin bounces off the front of the rim as in figures 4.3c or 4.4a. Since it is moving to the right, it can be described as having topspin. Suppose that the basketball goes on to hit the backboard. For simplicity, we assume that the velocity of the basketball is horizontal when it hits the glass; that is, the basketball goes straight into the backboard. Two things happen. First, the basketball is accelerated upward slightly. Second, the spin of the basketball is either slowed or reversed to CCW. Both effects occur because the force of friction on the basketball at the glass is upward. With a little luck or good management, the basketball will either go through the hoop or subsequently bounce back over the hoop.

Use of the Backboard: The Layup

We now consider intentional use of the backboard. The main use of the backboard in shooting is for layups. Layups have become more important over the past few years because of renewed interest in Pete Carril's Princeton offense.[2] One of the first things that I learned about the game is that, when driving to the basket or leading a fast break, it's better to shoot a layup by moving to the left or the right and bank in the basketball than to try to put it in directly over the front of the hoop. I must admit that I never understood why until now. These days, putting the basketball through the hoop by means of a finger roll with the hand over the hoop is popular. The following discussion does not apply to that. It only applies to ordinary shots from close-in. There are lots of players who, like me, do not have the option of a finger roll.

I used VideoPoint® to analyze video of the side view of a half-dozen right-hand layups. I was amazed by the wide range of ways that the basketball can bounce off the backboard and go in. The data for two extreme cases are shown in figure 4.7 where the angle that the velocity of the basketball makes with the horizontal versus time is plotted. The signs of the angles are strange and nonstandard. The angle above the horizontal is defined to be pos-

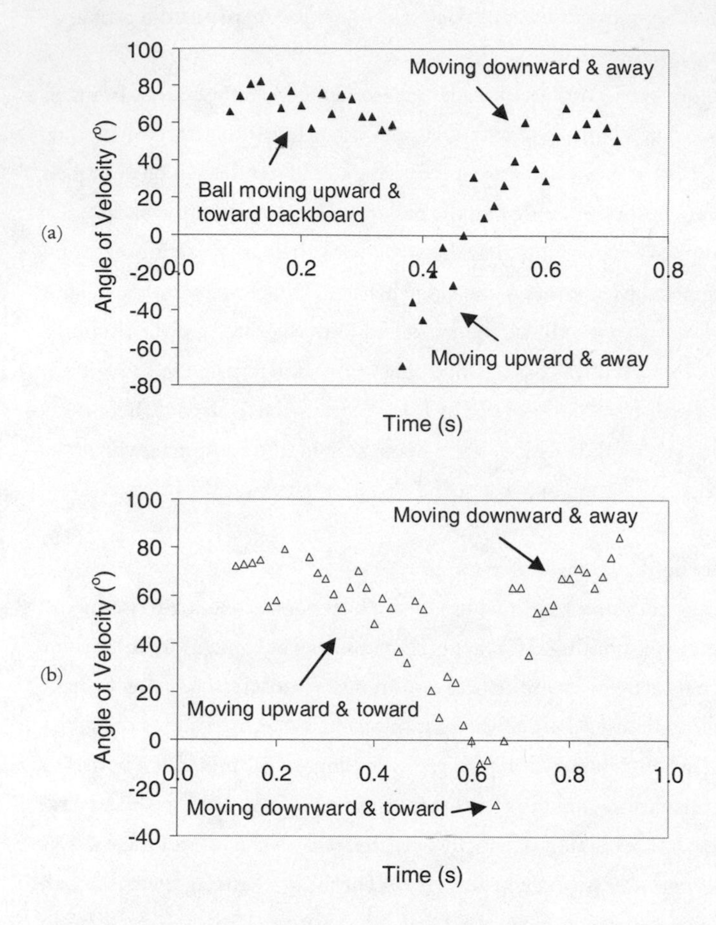

Figure 4.7. Angle that the velocity makes with the horizontal versus time for two different shots

itive if the ball is moving toward the backboard and negative if moving away. Defining the angles in this manner makes it easier to identify the collision.

Initially, for both shots, the angles are positive at about 70°. This indicates that the basketball is moving upward and toward the backboard at the point of release. In both cases the angle decreases as time increases. This indicates that the direction of travel of the ball is leveling off due to gravity as time increases.

For the layup represented in figure 4.7a, the ball is still moving upward when it collides with the backboard at about 0.37 s. Just after the collision with the backboard, the ball continues to move upward but away from the backboard. This is indicated by the negative angles labeled "Moving upward & away." The ball reaches the top of its arc at about 0.48 s. This is indicated by the angle of 0° at about 0.48 s. After 0.48 s, the ball begins to move downward and away from the backboard. The ball begins to enter the hoop at the end of the data, about 0.73 s.

The layup represented in figure 4.7b is different. The angle of the velocity decreases gradually to 0° at about 0.60 s. This implies that the ball reaches the top of its arc at about 0.60 s. The angle of the velocity then becomes negative until about 0.67 s when the basketball hits the backboard. The basketball is moving downward when it hits the backboard. Just after the collision with the backboard, the ball moves downward but away from the backboard. That is indicated by the positive angles after about 0.67 s. The ball begins to enter the hoop at the end of the data, about 0.93 s.

Figures 4.7a and 4.7b represent extreme circumstances. The shot represented in figure 4.7a is long and bounces in via the front of the rim. The shot represented in figure 4.7b is short and bounces in via the back of the rim. What is more common is for the ball to collide with the backboard while moving approximately horizontally then get nothing but net. The spin of the basketball is also important. I found that I shoot repetitive layups with a backspin of about 2 revolutions per second. This is about the same as for my outside shots. A basketball that has backspin when it hits the backboard usually experiences a downward force of friction. This helps accelerate the basketball downward. In addition, the force of friction usually slows the spin of the basketball. The direction of the spin does not change but the basketball is spinning more slowly after the collision with the backboard. Since the basketball is traveling in the opposite direction after the collision it can be described as having topspin after the collision with the backboard.

The most important result of the analysis was that, no matter how the ball hit the backboard, the speed of the ball bouncing away from the backboard (horizontal speed) was between 0.8 m/s and 1.3 m/s with an average of about 1.0 m/s (2.2 mph). Figure 4.8 shows the velocity of the basketball toward the backboard for the two extreme cases shown in figure 4.7. Negative values indicate the basketball is moving away from the backboard. The average horizontal speed away from and toward the backboard is approximately the same for both shots at about 1.0 m/s.

The observed range of speeds, 0.8 m/s to 1.3 m/s, is reminiscent of the range of speeds associated with figure 2.4. The analysis associated with that figure showed that a basketball that has an angle of approach of 0° and is 0.1 m above the top of the rim gets nothing but net for speeds between 0.62 m/s and 1.5 m/s. The similarity between the two ranges of speeds is reasonable. Both refer to the horizontal speed of a basketball approaching a hoop. For the layup, the basketball approaches the hoop traveling away from the backboard. For the straight-in shot, it approaches the hoop traveling toward the backboard. Whether a basketball with a horizontal speed of 1 m/s goes through the hoop after it bounces off the backboard depends on its vertical velocity. It also depends on the horizontal velocity along the backboard, but we will worry about that later. If the ball bounces upward, it could easily go beyond the front of the hoop. The reason is that it would spend more time in the air. In that case, the horizontal speed might cause the basketball to go beyond the front of the hoop. If the ball bounces downward, it could hit the back of the hoop.

Because of what we know about restrictions imposed by physics and because of the interrelationship between the horizontal and vertical velocities, it might appear that it is more difficult to make a shot off the backboard. However, what is hidden is that, because of friction and spin, most vertical and horizontal velocities are achievable when the ball bounces off the backboard. That is not the case for a basketball that is shot directly at the basket. As discussed in chapter 3, the laws of physics put severe restrictions on the path of a basketball in flight. Those restrictions are significantly reduced by the interaction of the basketball with the backboard. The same restrictions apply to the path of the ball in flight both before and after the bounce off the backboard, but the bounce disconnects the two

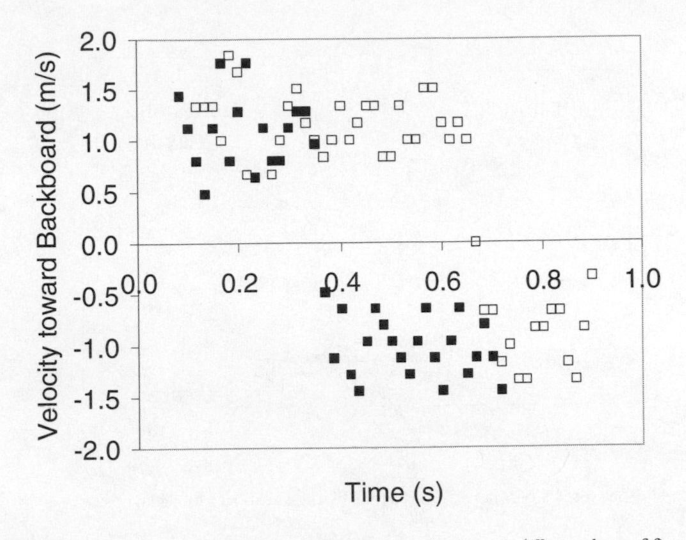

Figure 4.8. The velocity toward the basket versus time for the two different shots of figure 4.7

segments of the path. This occurs because additional forces act on the basketball during the bounce off the backboard.

In addition, there is very little dependence on the vertical velocity. I realized that after making the following calculation. I considered a basketball that bounced off the backboard 0.35 m above the hoop. The basketball was given a horizontal component of the velocity of 1 m/s after bouncing off the backboard. It was found that if the angle at which the basketball moves away from the backboard is greater than 57.3° below the horizontal (+57.3° in the notation of figure 4.7), the ball hits the back of the hoop. If the angle is greater than 33.9° above the horizontal (–33.9°), the ball goes beyond the front of the hoop. This very large range of allowed angles is the first reason why a layup is a high-percentage shot. The velocity of the basketball in the vertical direction doesn't matter much. So long as the horizontal velocity of the basketball when it comes off the backboard is about 1.0 m/s, it will go in.

I let the basketball hit the backboard 0.35 m above the hoop because that appears to be proper placement of the ball. The six layups that I analyzed hit the backboard at between 0.23 and 0.48 m with an average of

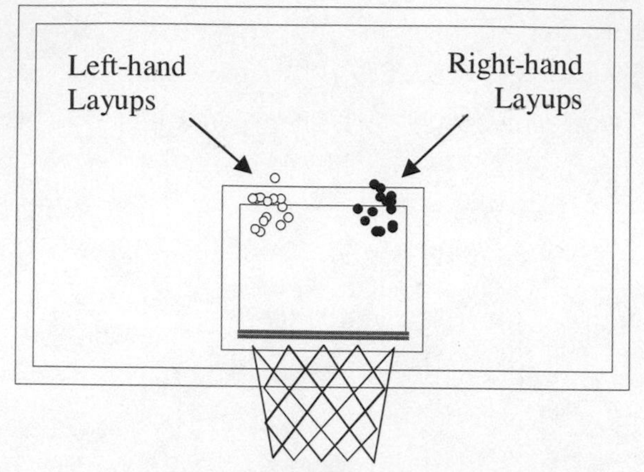

Figure 4.9. Scatter plot of where the basketball hit the backboard for 28 layups

0.35 m above the top of the hoop. In addition, I shot 14 left-hand layups and 14 other right-hand layups and analyzed videos from the front of the basket. Figure 4.9 shows where those layups hit the backboard. The average of the vertical positions of the 28 layups is just slightly below the bottom of the strip at the top of the rectangle painted on the backboard just above the hoop. According to International Basketball Federation (FIBA) specifications, the bottom of the strip at the top of the rectangle should be 0.35 m above the hoop. It appears that the designers of the backboard have given players a good target for the height at which a layup should hit the backboard.

The data also indicate the best horizontal position for the basketball when it hits the backboard. The center of the spots is 0.16 m (6.5 in) to the left of center for layups from the left and 0.16 m to the right for layups from the right. That the positions are different for left- and right-hand layups makes sense. The basketball continues to move horizontally after hitting the backboard. This suggests that a good training aid for teaching young players to shoot layups would be to put spots (e.g., tape) on the backboard along the bottom of the strip at the top of the rectangle 0.16 m to the left and right of center.

In the 2006 NCAA women's Final Four semifinal win over UNC Crystal Langhorne and Laura Harper of the University of Maryland Terrapins gave a clinic on the proper use of the backboard when shooting layups. Time after time they broke through the chaos in the paint to bank home a shot. I am almost certain that a plot of where those basketballs hit the glass would look just like figure 4.9. In many ways I'd rather watch a good women's basketball game than a good men's game. Strange as it may seem, what reduces my enjoyment of the men's game is the slam dunk. The slam dunk is indeed spectacular. The members of Phi Slamma Jamma are magnificent, talented athletes. However, the level of skill involved in a slam dunk is minimal. I would like to see the points awarded for a slam dunk reduced to one. Better yet, the height of the basket should be raised to 12 feet. I am not sure if many fans would agree with either suggestion. I am sure, though, that everybody except fans of the Lady Blue Devils would agree that one of the best games in either the men's or women's 2006 NCAA tournament was the Lady Terps' win over Duke for the national championship. Maryland's steady comeback (from a 13-point deficit) through the second half punctuated by Kristi Toliver's 3-point shot to tie the game with 6.1 s left was basketball at its best. The clutch play of Shay Doron, Marissa Coleman, and the rest of the team in overtime was as good as it gets. And then there is the coach, Brenda Frese. I don't think that I have ever seen a better presence on the sideline. Every TV shot of her during the second half when her team was losing showed her smiling with confidence. That had to have been an inspiration to her team. Since the starting five will all be back for the 2006–2007 season, it will be interesting to see whether it is the time of the turtle in women's college basketball. It's always easier to get into first place than to stay there. The Lady Terps are a welcome addition to high-level Division I local college basketball. When I first arrived in the Baltimore-Washington area it was Georgetown in the Big East and Maryland in the Atlantic Coast Conference. Then there were the David Robinson and Don Devoe eras of Navy basketball. This year it was the Lady Terps and the great run by the George Mason University Patriots in the men's NCAA tournament. There are numerous other colleges and universities in the area and every year someone emerges with a great team. Recently there has been a resurgence of the Wizards. Add all of that to the excellent basketball played by high schools such as Dunbar and De Matha and the result is a basketball lover's paradise. It's no wonder that John Feinstein, well-known fan, commentator, and author of books about basketball (*The Last Amateurs, A March to Madness,* and so on— the two books mentioned here were published by Little Brown and Company in 2000 and 1998, respectively), has a home in this area.

Back to the physics—in addition to very little dependence of the layup on the vertical velocity of the basketball, there is very little dependence on where the ball can hit the backboard and still go through the hoop. I realized that after making some calculations where the ball leaves the backboard horizontally at 1.0 m/s (2.2 mph) at different contact points. If the basketball hits the backboard at between 0.14 m (5.5 in) and 0.59 m (23 in) above the rim, it goes in. The range from 0.14 m to 0.59 m above the rim can be thought of as a vertical window associated with the backboard. This very large window is the second reason why a layup off the backboard is a high-percentage shot.

This discussion points out the essence of a standard layup. During the layup the ball is gently directed toward a spot on the backboard. The spots for right- and left-hand layups are located along the bottom of the strip at the top of the rectangle painted on the backboard 0.16 m to the right and left from the center. If the basketball approaches the backboard with a speed of about 1.0 m/s, there is a very high probability that it will go through the hoop. An example of good technique for shooting layups is shown in figure 4.10. Notice Shannon Johnson's eyes. They seem to be focused on the correct spot on the painted rectangle. Also, she appears to be shooting the basketball directly upward relative to her. That works perfectly if she is moving forward with a speed of about 1.0 m/s. If she is moving forward at 1.0 m/s, the basketball already has that velocity so there is no need for her to shoot the basketball toward the backboard. This is a very good technique for shooting layups. If a player is traveling at high speed toward the basket, as on a fast break, whenever possible she or he should brake to about 1.0 m/s, then softly project the basketball straight up.

Figure 4.10. Shannon Johnson #14 of the Connecticut Sun shoots a layup during the 2003 WNBA East/West All-Star Game at Madison Square Garden on July 12, 2003. ©2006 Jesse D. Garrabrant/WNBA Enterprises/Getty Images

Thus far, we have only seen why a layup is an easy shot. To show why it is better than a direct shot, we consider a direct shot from the same height as a foul shot (2.4 m) and a horizontal distance of 0.66 m (2 ft) from the front of the hoop. We use 0.66 m because the layups were shot from a horizontal distance of about 0.33 m (1 ft) from the hoop. Because the ball travels to the backboard then to the hoop the equivalent distance for a direct shot was taken to be twice as great as the distance from the ball to the backboard. This direct shot was analyzed in the same way that the foul shots were analyzed. For the direct shot from 0.66 m, the launch angle that minimizes the approach speed is 72° and the corresponding launch speed is 4.45 m/s. In addition, the minimum angle of approach to get nothing but net is –3.5° for a men's basketball and –4.5° for a women's basketball. The negative sign indicates that the basketball is on the way up when it is above the top of the front of the rim. However, we have already seen layups that get nothing but net for much more negative angles of approach. For example, the layup represented in figure 4.7a was traveling upward away from the backboard at an angle of about –40°. This is equivalent to an angle of approach of –40°. Such angles of approach are impossible for a direct shot from 0.66 m. Next, for a direct shot, the extreme positions of the center of the ball when it is above the rim for an angle of approach of 0° are 0.12 m and 0.16 m. This is a very small range of values when compared with the range of 0.14 m to 0.59 m for a layup where the basketball comes directly off the backboard with a speed of 1.0 m/s. Finally, the horizontal approach speed was fixed at 1.0 m/s for the short direct shot. The corresponding range of angles of approach for the close-in direct shot to get nothing but net was found to be 40.2° to 62.1°. This is much narrower than the range of 57.3° to –33.9° for layups with a fixed horizontal velocity of 1.0 m/s. All of this shows why it is much easier to make a layup than to just shoot the ball in directly over the front of the rim in this situation.

For completeness, we will compare the results for the close-in direct shot with those for a foul shot. We saw that the value of the launch angle that minimizes the approach speed for a foul shot is 50.8° and the corresponding launch speed is 6.75 m/s. The corresponding values for the close-in direct shot are 72° and 4.45 m/s. The values show that a shot from close to the basket should be launched at a higher angle and lower speed than for a foul shot. The angle of approach and minimum approach speed were found to be 27° and 1.44 m/s for the close-in direct shot. Those for a foul shot are 40° and 5.0 m/s. This shows that the close-in direct shot must approach the basket at a lower angle and with a much smaller approach speed than a foul shot. Finally, the vertical size of the window was calculated. At an approach angle of 40°, the vertical window for the close-in direct shot was found to be 0.83 m. The size of the window at the angle of approach for minimum approach speed, 27°, for the close-in direct shot was found to be 0.64 m. Both values are much larger than the value of 0.34 m for the foul shot. These results show why it is easier to make a close-in direct shot than it is to make a foul shot.

Finally, we consider further the effects of spin on the bounce from the backboard, in part, because while watching a Wizards-Mavericks game I heard Phil Chenier say that he had just watched a player "spin it in." To check this out, I made some videos while trying to spin in a basketball off the board. One thing that I learned from this exercise is that it's tough to put special spin on the basketball and still make it go in the basket off the backboard. The percentage of shots that I made was very low.

The front view of the trajectory of one of the successful shots is shown by squares representing the left-hand spin shot in figure 4.11. The points represent the position of the ball every 1/60 s. The point where the ball contacted the backboard is shown by the filled square.

The left-hand spin shot had spin that is CCW as viewed from above. That kind of shot would be useful when driving the baseline from right

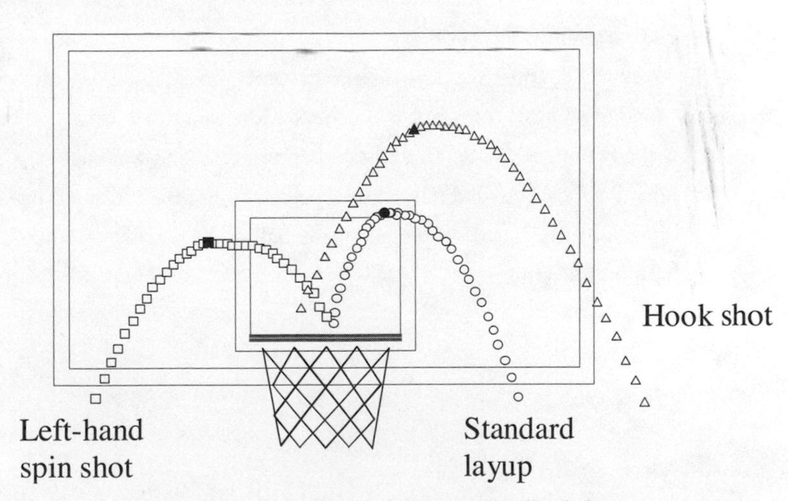

Figure 4.11. Front view of the trajectories of various types of bank shots

to left, then emerging at the left-hand side of the basket. For comparison, the front view of the trajectory for a standard right-hand layup is also shown in figure 4.11. For both the left-hand spin shot and the standard layup the point where the ball contacted the backboard is shown by the filled shape. Based on the trajectory for the standard right-hand layup, we conclude that the left-hand spin shot has a significantly different trajectory than a standard left-hand layup. The extra spin enables the shot to be banked in using a contact point both below and to the left of the usual contact point for left-hand layups. That is because the extra force of friction is to the right. That extra force gives the basketball more velocity to the right. Consequently, not as much height is necessary to get the basketball over the rim and into the basket.

Finally, the front view of the trajectory for a hook shot is shown in figure 4.11. The contact point for the hook shot is well above and to the right of the usual contact point for right-hand layups. On the basis of the relatively symmetrical trajectory, we conclude that the effect of spin and hence friction are small for this shot. This is a very useful shot for height-challenged players. Since the basketball is shot high on the backboard, it represents a way to successfully shoot the ball over a tall person. Hook shots are not used much these days, but more standard shots high off the glass are still widely used.

In this chapter, we have discussed some consequences of the bounce of a perfect basketball, but we haven't addressed the issue of why a basketball bounces. We do that in the next chapter. That will enable us to see how a real basketball differs from a perfect basketball. Along the way we consider some actual bounces of the ball and see what an impact they have on the game.

That's the Way the Ball Bounces

In early 2005, my son and I watched a Wizards versus Mavericks game on TV.[1] We noted that the ball was dribbled by the players about 1,593 times while the clock was running. The breakdown by quarter was 413, 466, 366, and 348 bounces. It would be interesting to know if the coaches said anything at half-time about too much dribbling. Certainly, the game flowed better in the second half. The number of dribbles by players at the foul line was 280, and we counted 60 bounce passes. This does not include bounces of the ball by the referees who seem to like to dribble and use the bounce pass extensively. There were 301 miscellaneous bounces of the ball where the ball bounced off the floor during a steal, after going through the hoop, and so on. Also, the ball bounced off the backboard 102 times, bounced off the rim and ultimately into the basket 89 times and bounced out of the hoop 129 times. The grand total of the number of bounces that we counted is 2,494 and I'm sure that we missed quite a few. This suggests that another factor that determines a player's level of success at the game is her or his ability to control the bounce of the basketball.

It's probably not generally known or appreciated that it is the height of the bounce that determines the proper inflation pressure for a basketball. One of FIBA's rules is that "the ball shall be inflated to an air pressure so that when it is dropped onto the playing surface from a height of approximately 1.8 m, measured from the bottom of the ball, it will rebound to a height of between 1.2 m and 1.4 m measured to the top of the ball." Consequently, every gym should have a wall with marks on it at heights of 1.2, 1.4, and 1.8 m.

Because the bounce of the ball is so important, it's worth spending more time discussing it. Warning: From here until the discussion of figure 5.10 the chapter is physics-oriented and some of it's a bit "teachy." If you really hate physics, you might prefer to skip to the vicinity of figure 5.10.

We have already dealt with the bounce of a basketball from the rim and the backboard. We have seen how contact forces, known as the normal force and the friction force, control the bounce. The normal force and the contact friction force act on the outside of the ball. There are also forces on the basketball caused by the air pressure on both the inside and outside of the basketball. Finally, there are an elastic force and an internal friction force that act within the cover and rubber bladder of the ball. This gives us a grand total of nine forces. There are many ways of classifying the forces. Three of them (static friction, air drag, and internal friction) are frictional. (One of the purposes of this book is to separate fact from friction.) Four (the buoyant, Magnus, air-drag, and pressure forces) are caused by the air. In fact, all of the forces except gravity are caused by the electrical force. Gravity is a fundamental force. It is not currently thought to be caused by any other force.

Let's start by considering a nonspinning basketball that was in the air moving downward and is now in contact with the floor. What happens is that when a basketball collides with the floor, it pushes (down) on the floor. A physicist would say that the basketball exerts a normal force downward on the floor and we will refer to that force as $\mathbf{N}_{\text{ball on floor}}$. $\mathbf{N}_{\text{ball on floor}}$ does not affect the basketball because it is a force on the floor. Fortunately, there is also a force on the ball. We will refer to that force as $\mathbf{N}_{\text{floor on ball}}$ because it is the normal force of the floor on the ball. This $\mathbf{N}_{\text{floor on ball}}$ is the reaction force required by N3L and is what causes the basketball to bounce.

The introduction of two forces, $\mathbf{N}_{\text{ball on floor}}$ and $\mathbf{N}_{\text{floor on ball}}$, associated with the collision of the ball with the floor might seem overdone. Unfortunately, it's essential and is not just semantics. Part of the reason that it's essential is

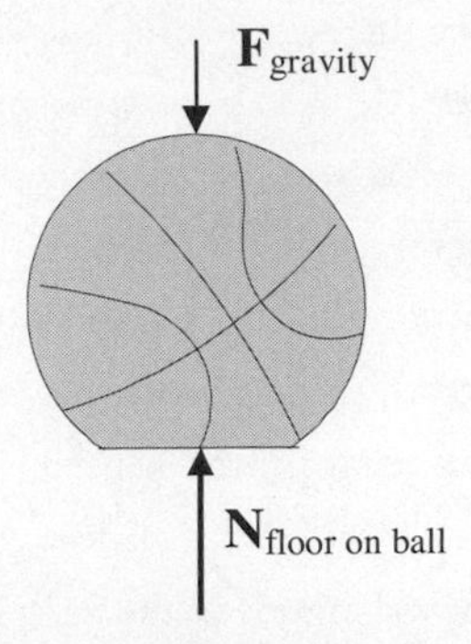

Figure 5.1. The primary external forces on a nonspinning basketball colliding with the floor

that these two forces, $\mathbf{N}_{\text{ball on floor}}$ and $\mathbf{N}_{\text{floor on ball}}$, act in opposite directions. They also act on different objects—one acts on the floor and the other acts on the ball. These are the action and reaction forces dictated by N3L. It's not unreasonable to think of $\mathbf{N}_{\text{ball on floor}}$ as the action and $\mathbf{N}_{\text{floor on ball}}$ as the reaction. Since the ball bounces, it is easy to see the effect of $\mathbf{N}_{\text{floor on ball}}$. Why don't we notice the effect of $\mathbf{N}_{\text{ball on floor}}$? The answer follows from the facts that the floor is connected to the building, which is connected to the Earth, the mass of which is huge. Since N2L tells us that the acceleration is inversely proportional to the mass (the larger the mass the smaller the acceleration), the acceleration of the floor-building-Earth system is miniscule, so we don't notice that the floor bounces away from the ball, but it does.

The primary external forces on a nonspinning basketball during a bounce are shown in figure 5.1. A model for the shape of the basketball is also shown. I was worried that the sketch in figure 5.1 wasn't realistic so I studied a basketball colliding with the floor. I found that the deformation of a basketball is small and difficult to see using my crude video equipment. I visited the local Toys "R" Us® store and bought a four-square ball. It is about the same size as a basketball and is much squishier. I took some videos of the four-square ball colliding with a Vernier Force Plate® that I borrowed from the USNA physics department. One of the frames of the video is shown in figure 5.2. The picture leads me to believe that figure 5.1 is a good representation of a basketball during its collision with the floor.

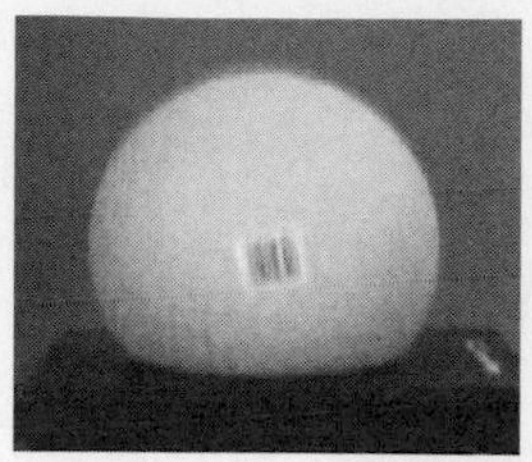

Figure 5.2. A four-square ball collides with a force plate

Next, we consider the results of some experiments that will enable us to learn something about $\mathbf{N}_{\text{floor on ball}}$. These days it is fairly easy to make measurements of the force on an object versus time. I borrowed a LabPro® computer interface and LoggerPro® software from the USNA physics department and used it with the Vernier Force Plate®. That made it possible to measure the magnitude of the force versus time that a dropped or dribbled basketball exerts on the force plate. The force plate was bolted to the horizontal surface of a sturdy desk. About 650 N (150 lbs) of weight were piled on the desk next to the force plate to reduce vibration. The data for the force on the force plate versus time for six bounces of a nonspinning men's basketball dropped from about 1.1 m (3.6 ft) are shown in figure 5.3. The times for the six bounces were shifted so that the data appear to fall on a single curve.

On the basis of N3L, the data in figure 5.3 also represent the magnitude of the force versus time for the force of the force plate on the basketball. The force increases rapidly over the first 0.008 s to a maximum value of about 650 N (150 lbs) then decreases rapidly to zero in about the next 0.008 s. The maximum force is large in that it is more than 100 times the weight of the basketball, which is 6.0 N (1.3 lbs). In addition, the ball is in contact with the force plate for a short time, about 0.016 s.

The large force confirms something that I have known for a long time—basketballs are dangerous. A few days after I was rewriting this section, my son broke two fingers while playing in a pickup game. He was hit by the basketball. He's strong but even his two fingers can't withstand a force of 650 N.

There is a way to test whether the data are reasonable—obey the laws of physics. Newton's Second Law tells us that the area under the curve (the shaded portion) in figure 5.3 should be equal to the mass of the basketball times the magnitude of its change in velocity. There is a special name that physicists give to the area under the curve in figure 5.2—

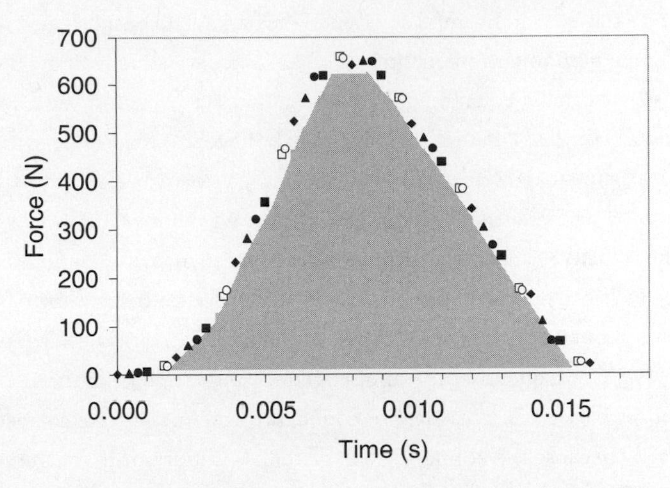

Figure 5.3. Force versus time that a nonspinning men's basketball dropped from 1.1 m exerts on a horizontal force plate

impulse. The general definition of impulse is given on the left-hand side of the following equation.

$$\text{Impulse} \equiv \int_{t_i}^{t_f} \mathbf{F}dt = m\mathbf{v}_f - m\mathbf{v}_i \equiv \text{change in momentum} \qquad (5.1)$$

The meaning of the three-line equal sign is "defined to be." Impulse, then, is given to a basketball when a force acts on it over a time interval (some initial time, t_i, to some final time, t_f). The impulse represented in figure 5.3 is due to the force of the floor on the ball so that the direction of the impulse is up. The magnitude of the impulse associated with figure 5.3 is about 5 N s. This is not quite equal to the total impulse on the basketball during the bounce. In addition to the upward impulse of the force plate on the ball there is a downward impulse of gravity on the ball. If we multiply the constant weight of the basketball times the time that the ball is in contact with the force plate, we get an impulse due to gravity of about 0.1 N s. Consequently, the total impulse on the basketball is about 4.9 N s up.

The mass times the velocity that appears on the right-hand side in equation (5.1) also has a special name—**momentum**. Consequently, equation (5.1) tells us that the impulse given to the basketball must be equal to the change in momentum of the basketball. Since we already know the impulse, we can use equation (5.1) to check our data. We know the mass of the basketball. In addition, a video of the bounce was made during the force experiment and was used to determine the velocity before and after the bounce. The velocity just before the bounce was about 4.5 m / s down and the velocity just after the bounce was about 3.5 m / s up. This gives a change in velocity of the ball of about 8 m / s up. Since the mass of the basketball is about 0.61 kg, the mass times the change in velocity is also about 4.9 N s up. Consequently, the impulse is equal to the change in momentum of the basketball. This gives us confidence that our data are correct.

Partly because the bounce of a basketball is important to the game and partly because something happens during the bounce that physics teachers sometimes forget about, it is worth analyzing the bounce more carefully.

1. First stage of the bounce. For a short time after the basketball contacts the floor, probably for less than 0.001 s, the ball continues to accelerate (increase its speed) downward even though it is in contact with the floor. The reason is that the upward normal force of the floor on the ball, $\mathbf{N}_{\text{floor on ball}}$, is initially very small. Consequently, just after the ball first contacts the floor, gravity dominates and the ball continues to accelerate downward. As time passes, $\mathbf{N}_{\text{floor on ball}}$ increases and the acceleration decreases. Eventually, a time is reached when the magnitude of $\mathbf{N}_{\text{floor on ball}}$ equals the strength of the downward force of gravity. At this first special time, the total force is zero and the acceleration is zero though the ball is still moving downward.
2. Second stage of the bounce. After the first special time, the magnitude of $\mathbf{N}_{\text{floor on ball}}$ is greater than the magnitude of the downward force of gravity—$\mathbf{N}_{\text{floor on ball}}$ dominates. In this case the acceleration is upward though the velocity is still downward. What is happening is that the speed of the basketball is decreasing. The ball continues to slow until the ball stops. The second special time is when the ball stops. For the graph

shown in figure 5.2, this second special time is about 0.008 s where $\mathbf{N}_{\text{floor on ball}}$ is a maximum. The ball has its maximum deformation at this time. Since $\mathbf{N}_{\text{floor on ball}}$ is a maximum, the upward acceleration is a maximum, but the velocity of the basketball is zero at this time. A very short time after stopping, the basketball begins moving upward, and the completion of the bounce just reverses the process.

3. Third stage of the bounce. The basketball increases its speed upward until the magnitude of the $\mathbf{N}_{\text{floor on ball}}$ equals the magnitude of the downward force of gravity.
4. Fourth stage of the bounce. The basketball slows because the magnitude of the downward force of gravity is greater than the magnitude of $\mathbf{N}_{\text{floor on ball}}$.

The basketball then loses contact with the floor. After the ball loses contact with the floor, the forces described in chapter 1 take over. (The drag and buoyant forces also act on the nonspinning basketball during the bounce. They have been ignored in this discussion because they are very small compared with $\mathbf{N}_{\text{floor on ball}}$.)

This explains the bounce in general terms but the very large maximum force exerted on the ball by the floor and short contact time are mysteries. To further analyze the bounce, we need to consider the forces caused by air pressure and the elastic and friction forces within the cover.

The pressure force is caused by collisions of air molecules with the surface. The force on a surface due to an individual air molecule is very small. However, usually there are billions and billions of air molecules so that the total pressure force on a surface can be large. The pressure inside a basketball is adjusted by changing the number of molecules inside. For example, an increase in pressure occurs when a pump puts extra air molecules into a basketball. The other effects of air that we have considered, the buoyant force, the drag force, and the Magnus force, are also caused by collisions of air molecules with a surface. They, too, can be explained by varying the numbers of molecules near a surface.

The strength of the pressure force on a surface, $F_{pressure}$, is calculated by multiplying the pressure, P, times the area, A. The equation is

$$F_{pressure} = PA. \tag{5.2}$$

We also need to keep in mind that the pressure force has a direction. The pressure force is perpendicular to and toward the surface that the air is in contact with. Consequently, for a calculation based on equation (5.2) to be valid, the area must be small enough so that the force is everywhere in the same direction and has the same strength. The metric units of pressure are newtons per square meter. A newton per square meter is known as a Pascal (Pa). The English units for pressure are the familiar pounds per square inch (psi) (1 psi = 6.9 kPa). It's written on most basketballs that the pressure inside the ball is supposed to be 7–9 psi above atmospheric pressure. This is sometimes referred to as gauge pressure. If the gauge pressure is 8 psi (55 kPa), the total pressure, sometimes referred to as absolute pressure, inside the ball is about 23 psi (160 kPa) since atmospheric pressure is about 15 psi (104 kPa).

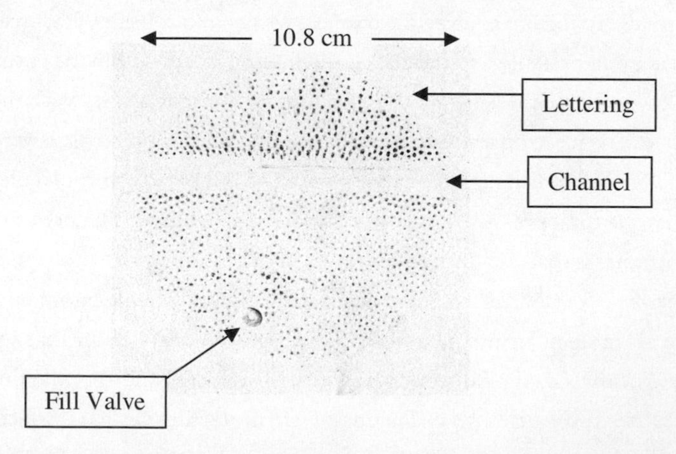

Figure 5.4. Imprint that a nonspinning men's basketball dropped from about 1.3 m made on carbon paper on a horizontal force plate

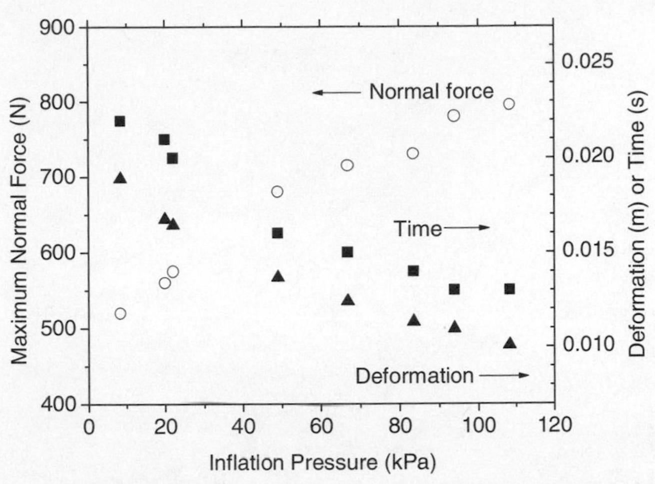

Figure 5.5. Maximum normal force (open circles), contact time (filled squares), and maximum deformation (filled triangles) versus inflation pressure during a bounce for a men's basketball dropped from 1.1 m onto a force plate. The inflation pressure of a typical basketball is 55 kPa.

To calculate the force on the flat bottom, we need the area. I dropped a properly inflated men's basketball from a height of about 1.3 m onto carbon paper. The carbon paper was carbon side down against plain paper that was on the force plate. A typical imprint made by the basketball during the bounce is shown in figure 5.4. The size of the imprint is relatively small showing that a basketball doesn't deform much during a bounce.

The maximum area where the basketball hit the paper had a diameter of about 10.8 cm and this corresponds to an area of 9.2×10^{-3} m^2. Using this area and a pressure of 160 kPa, we calculate using equation (5.2) that the strength of the downward pressure force on the inside of the flat bottom is 1,470 N (330 lbs). The pressure force is larger than the maximum strength of the upward normal force, $\mathbf{N}_{\text{floor on ball}}$, 650 N.

While I was at it, I varied the pressure in the basketball, then measured the force versus time and the imprint on carbon paper for a men's basketball dropped from 1.1 m onto a force plate. Figure 5.5 shows the results. The maximum normal force increases as pressure increases. This

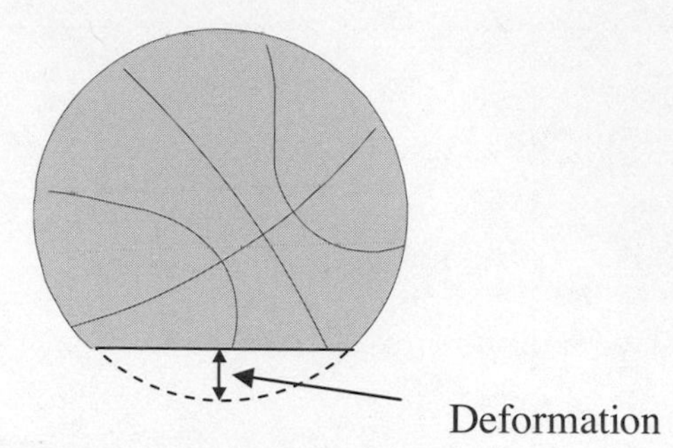

Figure 5.6. Definition of the deformation

makes sense. While getting hit by a basketball is never pleasant, I would rather get hit by a basketball with a small amount of air in it (low pressure) than by a basketball containing a lot of air (high pressure). The basketball containing a lot of air is harder or more rigid. It hurts more because the force that it is capable of exerting is larger. Also, the contact time decreases as pressure increases. Finally, the deformation of the basketball is plotted versus pressure in figure 5.5. The deformation is defined as the amount that the basketball is pushed in at the bottom. This is shown in figure 5.6. The deformation was calculated using the size of the imprint and the radius of the basketball. Figure 5.5 shows that the deformation of the basketball is small and decreases as pressure increases.

One way to get further insight into the bounce is to model a basketball by replacing the basketball with an equivalent spring and mass. This should work reasonably well because a spring has many of the characteristics of a basketball. If we push on a spring, it pushes back. If we place a mass on top of the spring, then drop the spring and mass onto the floor, it bounces. A familiar example of a bouncing spring and mass is a person bouncing on a pogo stick.

The primary reason that we do this is that we can write equations that describe the behavior of a spring and mass. It is a topic that is addressed in all general physics courses. The basic equation describing what a spring does is

$$F_{spring} = -kx, \tag{5.3}$$

where x is the amount that the spring is compressed and F_{spring} is the force of the spring on the mass. Equation (5.3) is known as Hooke's law. It tells us that the more a spring is compressed, the greater the force. If we double the compression, the force doubles, and so on. The stiffness of the spring is given by the spring constant, k. A large value of k is associated with small compressions or large forces. The minus sign tells us that if a spring is compressed, it pushes back. If a mass is connected to the top of a spring connected to the floor and is given a shove up or down, the mass oscillates. This motion is usually called simple harmonic motion (SHM). The equation for the period of oscillation (time for one cycle) of the mass and (massless) spring is

$$T = 2\pi\sqrt{\frac{m}{k}} \tag{5.4}$$

Equations (5.3) and (5.4) contain force, time, and compression—just what we need to describe the data for a basketball. If we have the equivalent spring constant for a basketball we can describe the data.

To determine the equivalent spring constant for a basketball, we'll use the deformation and force data. We'll let the deformation of the basketball be equal to the compression of the equivalent spring. We know that the maximum deformation of a properly inflated basketball is about 0.013 m. We also know that the corresponding maximum force that the basketball exerts is 650 N. Plugging these values into equation (5.3) for x and F_{spring} we get an equivalent spring constant, k, of about 50,000 N/m. We can now use the model to predict the contact time for the basketball. If we plug k and the mass of the basketball into equation (5.4), we get 0.022 s. The predicted contact time of a basketball is about half this or 0.011 s. It is half because the down and back motion of a basketball during the bounce constitutes roughly half of a complete cycle. The contact time from the model, 0.011 s, is the same order of magnitude as the experimental value, 0.016 s. Considering the simplicity of the model, the result isn't all that bad.

The model gives us a good way to think about the bounce. A properly inflated basketball behaves as though it is a stiff spring. A small deformation gives rise to a large force. In addition, the contact time must be very short because the mass is relatively small. The model also enables us to explain other features of the bounce of a basketball. In figure 5.5 we see that the force increases and the deformation decreases as the inflation pressure increases. This implies that the stiffness of a basketball, the effective spring constant, increases as the inflation pressure increases. This predicts that the contact time should decrease as the inflation pressure increases, and it does.

Until now everything that we have said about the bounce is superficial. We have not considered what it is about the basketball itself that enables it to behave the way it does. To get some insight into that, we need to dissect it. What we'll do is separate a basketball into three sections: the upper hemisphere, the lower hemisphere minus the flat portion, and the flat portion.[2] Those sections and the forces on them are shown in figure 5.7. It is assumed that the deformation is large enough so that the basketball is accelerating upward. For clarity, the weight of (force of gravity on) each part has been omitted. Also, only the total pressure force is shown on each section. The total pressure force is the pressure force on the inside minus the pressure force on the outside.

From this point of view, the upper hemisphere is pushed upward by the pressure force. It is also pulled downward by an elastic force due to the truncated lower hemisphere. The upward acceleration of the upper hemisphere is equal to the pressure force minus the elastic force minus the weight divided by its mass.

The truncated lower hemisphere is pulled upward by the elastic reaction force to the downward elastic force on the upper hemisphere. There are three downward forces on the truncated lower hemisphere: the pressure force, the elastic force caused by the flat bottom, and its weight. Since it is inclined at an angle, the pressure force is only partially effective in pushing the truncated lower hemisphere downward. The downward elastic force is only partially effective for the same reason. The downward elastic force is weaker still because the length over which it acts is smaller than the length over which the upward elastic force acts. This assumes that the elastic force per length is approximately constant along the boundaries of the sections. The weight of the truncated lower hemisphere is small.

There are two upward forces on the flat bottom—the upward normal force that we have seen previously and the upward elastic reaction force to the downward elastic force on the truncated lower hemisphere. There are two downward forces on the flat bottom—a downward pressure force and its weight.

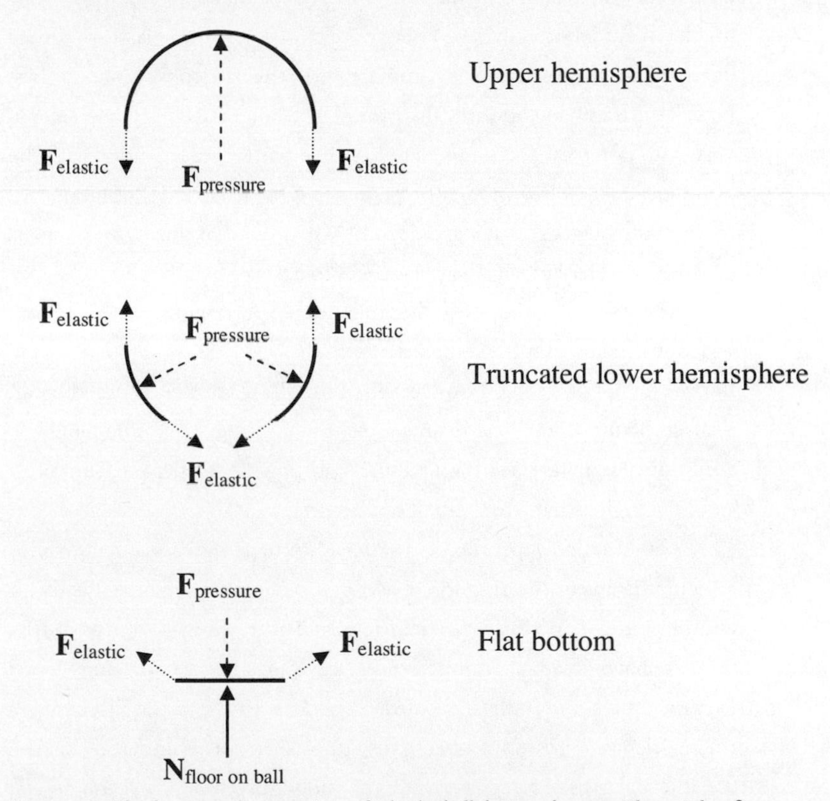

Figure 5.7. The forces on three sections of a basketball during a bounce. The weight of each part (the downward force of gravity) has been omitted and only the total pressure forces are shown.

Dissecting the basketball in this way allows us to see the roles of the pressure or the elastic forces in the bounce. The special feature of the elastic force is that it depends on the properties of the material. A basketball and a four-square ball bounce differently because they are made of different materials. The different materials give rise to different elastic forces. The result is different contact times, deformations, and maximum forces.

A problem exists with the models discussed so far. If we ignore air resistance, they predict that a basketball will always bounce to the same height that it is dropped from. When the basketball first contacts the floor, it is said to have potential energy (energy of position/height) and kinetic energy. For a perfect basketball and no air resistance, as the ball deforms while moving downward, both kinds of energy are transformed to elastic potential energy. The transformation is complete when the ball stops. When the basketball starts moving upward, all of the elastic potential energy then reconverts into potential and kinetic energy. The ball leaves the floor with the same speed that it had when it first contacted the floor and thus it bounces to the same height that it started from. We know that this never happens because a real basketball always bounces to a lower height.

To make this quantitative, the following experiment was carried out. Video was taken of a basketball that was dropped 1.8 m onto a concrete floor. The height of the bounce above the floor was determined by an analysis of the video. The results are shown in figure 5.8. The height of the bounce is always less than 1.8 m. Also, the higher the inflation pressure, the higher that a basketball bounces.[3]

As discussed in appendix VI, a small amount of the loss of height after a bounce is due to the drag force. Most of the loss of height is due to yet another kind of friction, internal friction. There are many kinds of friction. We have already mentioned three: air drag, static friction, and kinetic friction. In a general physics course, another kind is often discussed—rolling friction. Those four forces plus internal friction make a total of five.

Internal friction forces act when the rubber or leather molecules move relative to one another. This happens while a basketball is deforming, that is, when it is changing size or shape. The internal friction force behaves

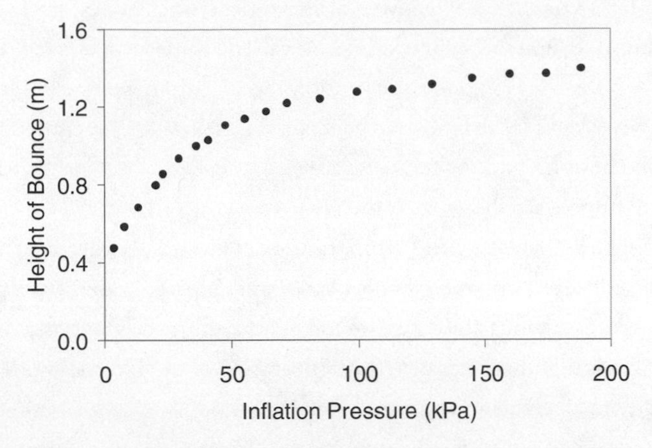

Figure 5.8. Height of the bounce versus inflation pressure for a basketball dropped from a height of 1.8 m onto a concrete surface

similarly to the kinetic friction forces involved in starting a fire by rubbing two sticks together. The kinetic friction force also acts when a crate slides across a floor. If a crate is given a push so that it slides on a horizontal surface, it always slows down because of the kinetic friction force. The temperature of the crate rises. One way to describe these processes is to say that the work being done on the sticks in starting the fire or the kinetic energy (energy of motion) of the crate is transformed to internal energy by the kinetic friction force. In both cases, the internal energy manifests itself mostly as a rise in temperature.

To show that the internal friction force has a great deal in common with the kinetic friction force, I measured the temperature of a dribbled basketball. I borrowed a thermometer with a resolution of 0.1°C (0.18°F) from the USNA physics department. The temperature of the gymnasium (air and floor) was measured to be constant between 20.0°C and 20.1°C. Measurements of the temperature of the basketball were made by inserting the sensor into the air valve opening. The ball was dribbled vigorously. Gloves were worn to eliminate contact between the hand and the ball. After the dribbles, the basketball was placed on the floor and the temperature was measured. The sensor was left in the basketball for times varying from 30 s to 120 s. This allowed the basketball to achieve thermal equilibrium, at least in the vicinity of the air valve. The process was repeated several times. During one equilibration the thermometer was inserted into the air valve three times. No effect on the observed temperature was found. The results for the temperature of the basketball versus the number of bounces are shown in figure 5.9.

Dribbling increased the temperature of the basketball up to 1°C (1.8°F). All of the observed temperatures were greater than the temperature of the gymnasium. The initial temperature of the basketball, 20.2°C, was slightly higher than the temperature of the gymnasium. That was a residual temperature increase from a previous dribbling experiment. The temperature rise is attributed to the internal friction force. The increase in the temperature of a basketball due to dribbling gives new meaning to warming up for a game.

It follows that when the internal friction force acts during a bounce, some energy is transformed to internal energy. Because of the internal friction force, some of the initial potential and kinetic energies are transformed (or lost) to internal energy. Consequently, when the basketball stops at the bottom of the bounce, the elastic energy is less than the sum of the initial potential and kinetic energies. Further energy is lost as the

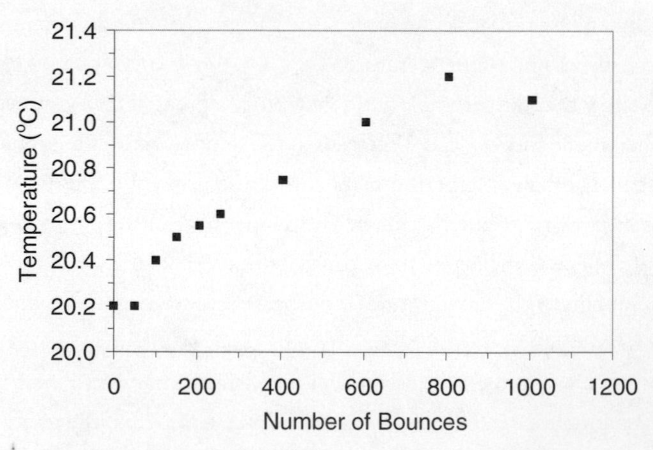

Figure 5.9. Temperature of a basketball versus the number of bounces

basketball moves upward. As a consequence, the ball leaves the floor with a smaller speed than it had when it first contacted the floor. Thus, a basketball bounces to a lesser height than it is dropped from. This also explains why a basketball that contains less air (lower inflation pressure) doesn't bounce as high. A ball that has less air in it deforms more so that the internal friction force has more effect.

That's about as much as I know about how a ball bounces. Admittedly, the discussion so far in this chapter has been physics oriented. Very shortly, we'll get back to some basketball and end this chapter with a discussion of dribbling and the bounce pass. Before we do that, however, we need to describe static and kinetic friction in some detail. We have been postponing this discussion since chapter 3. It's not too late because it sets up our discussion of dribbling and the bounce pass.

As mentioned a few times previously, (external) friction forces act parallel to surfaces in contact. The friction force is static if there is no slipping. Static friction opposes the tendency for relative motion of the touching surfaces. For example, suppose that we ask what it is that enables a person to walk on a horizontal floor. When a person at rest tries to walk forward, she or he attempts to push backward on the floor. If the floor (or shoes) were frictionless, the person's foot would slip backward along the floor. The person would remain at rest. A force in the forward direction is required to keep the person's foot from slipping and to enable the person to accelerate. That force is the static friction force. From a N3L point of view, when trying to walk forward, a person pushes back on the floor so that the floor pushes forward on the person. The person then accelerates forward.

Static friction is responsible for the ability of a person to maneuver on a basketball floor without slipping. It is usually the static friction force to the left that enables a person to cut (make a quick move) left, and so on. Figure 5.10 shows the static friction force that enables Becky Hammon to accelerate to the left.

Static friction has some weird characteristics. A floor only exerts as much static friction force on Becky as it has to to avoid slipping. If Becky pushes on the floor to the right with 1 N, the floor pushes to the left with 1 N. If she

Figure 5.10. Becky Hammon #25 of the New York Liberty drives (accelerates) around Raffaella Masciadri #33 and Lisa Leslie #9 of the Los Angeles Sparks during the WNBA game at Staples Center on July 5, 2005 in Los Angeles, California. ©2006 Andrew D. Bernstein / WNBA Enterprises / Getty Images

pushes with 2 N, the floor pushes with 2 N, and so on. There is an upper limit to the static friction force. If Becky tries to push on the floor with a force greater than the maximum force that static friction is capable of providing, her foot slips to the right. These facts are summarized by the following equation that gives the magnitude of the static friction force

$$F_{\text{static friction}} \leq \mu_{\text{static friction}} N \tag{5.5}$$

$\mu_{\text{static friction}}$ is known as the coefficient of static friction and depends on the nature of the surfaces in contact. N is the magnitude of the normal force of the floor on the shoe. If Becky's foot slips, kinetic friction takes over. The force of kinetic friction is still to the left but it is weaker than the maximum force of static friction. The equation that describes the strength of the kinetic friction force is

$$F_{\text{kinetic friction}} = \mu_{\text{kinetic friction}} N \tag{5.6}$$

$\mu_{\text{kinetic friction}}$ is known as the coefficient of kinetic friction and also depends on the nature of the surfaces in contact. Equation (5.6) says that the kinetic friction force has only one value. It would be the same regardless of Becky's speed. It is usually found that $\mu_{\text{kinetic friction}} < \mu_{\text{static friction}}$, which describes the fact that the kinetic friction force is usually smaller than the maximum static friction force. This implies that kinetic friction provides a smaller acceleration than the maximum static friction force.

We know that both static and kinetic friction act on basketball shoes. There is no slipping for a great deal of the time, but if you listen closely to a practice or a game, you will hear a great deal of intermittent squeaking. That noise occurs when a shoe slips. The squeak is caused by the kinetic friction force and usually occurs when a player starts, stops, or cuts hard. The force required for a player to start or stop fast or cut hard is often larger than the maximum force that static friction is capable of providing. The consequence is that some slipping and hence squeaking occurs. The kinetic friction force brings the foot to a stop after which static friction takes over.

It would seem that the greater force provided by the static friction force would make shoe design easy. It might be concluded that all that is required would be to make a shoe sole that would prevent slipping when the shoe is in contact with a basketball floor. This would entail making $\mu_{\text{static friction}}$ as large as possible. I am painfully aware, however, that this is not a good idea. We were practicing during the Christmas break of my freshman year in college. The floor had just been varnished and was capable of providing a very large force on a shoe before slipping occurred. Unfortunately, the maximum static friction force that the floor was capable of providing was too large. The medial meniscus in my left knee wasn't able to provide enough force to enable the lower leg to turn the foot and ultimately the shoe, that is, my upper body turned but my left foot and lower leg didn't. I ended up with a torn cartilage. One critical factor in basketball shoe design, then, is that a player must be able to pivot (turn on one foot) fairly easily. This requires that the shoe be able to slip rotationally fairly easily. The squeaking sounds of the game imply that shoes are also designed to allow some translational slipping for the average player. Designing shoes with the optimal values of $\mu_{\text{static friction}}$ and $\mu_{\text{kinetic friction}}$ must be an ongoing concern of the shoe companies. What is difficult is that $\mu_{\text{static friction}}$ and $\mu_{\text{kinetic friction}}$ depend on the nature of the surfaces in contact. The same shoe will behave differently depending on the surface. Consequently, it is impossible to design one shoe that works well on all surfaces. I've noticed that different shoes are now available for different surfaces. The physics justifies that.

There are many other considerations in shoe design. The amount of force that a basketball shoe experiences is often very large and thus shoes must be very strong. I was practicing with a Westminster teammate, Mike Drespling, at New Castle High School one summer when he made a cut. One of his Converse Chuck Taylor® All Stars® ripped in half. The rip was vertical so that the front separated from the back. By coincidence I met Chuck Taylor in the spring of 1966.[4] With my usual lack of diplomacy, I asked him about the shoe that failed and he said that it happens sometimes because they are made of canvas. I must admit that I only used a couple of pairs of Converse All Stars® though I am wearing them in figure 3.8. I mainly used Pro-Keds®. It's interesting that both All Stars® and Pro-Keds® are still available despite the large number of types of shoes that are now on the market. In fact, Converse All Stars® have become quite fashionable worldwide as a leisure shoe.

Let's now reconsider the effect of static friction on the bounce of a basketball. That was briefly discussed in chapter 4 and applied to bounces of a basketball from the rim and backboard. Here we'll consider the effect of static friction in more detail and apply it to dribbling and the bounce pass.

Suppose that a falling ball is spinning CCW when it hits the floor as shown in figure 5.11. The bottom of the basketball is moving both downward and to the right when it first contacts the floor. Because of the downward velocity, the ball exerts a downward force on the floor. As discussed in great detail earlier in this chapter, this requires that an upward normal force of the floor be exerted on the ball, $\mathbf{N}_{\text{floor on ball}}$. This is the force that makes the basketball bounce upward. Because of the velocity of its surface to the

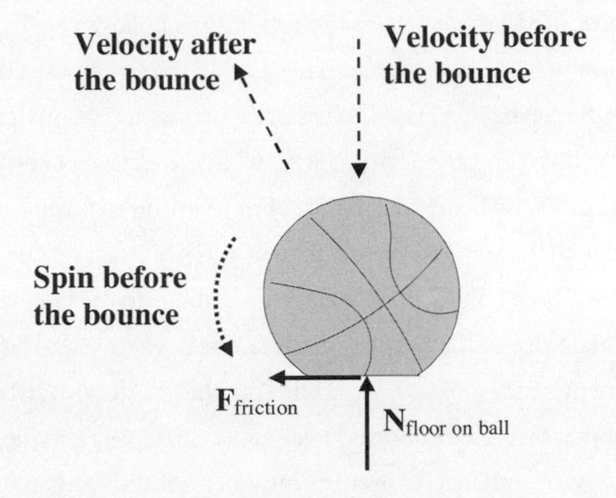

Figure 5.11. The contact forces on a spinning basketball touching a surface

right, the basketball exerts a force on the floor to the right. We don't care about that force since it is a force on the floor. The important force is the floor pushing back on the ball to the left and this is shown as $\mathbf{F}_{friction}$. If the ball does not slip when it contacts the floor, $\mathbf{F}_{friction}$ is a static friction force. If the ball is spinning extremely rapidly, the ball will slip so that the force of friction is kinetic rather than static. This discussion of the contact forces of a basketball with a floor is simplistic. Recent work by Cross shows that the forces of contact between a spinning ball and a surface can be quite complicated.[5] Grip-slip behavior is observed. However, our approach correctly accounts for the general features of the bounce described in this book.

Two things happen because of $\mathbf{F}_{friction}$. The first is familiar to anyone who has bounced a spinning basketball. The ball has a component of velocity to the left when it leaves the floor. That is because the ball accelerates to the left because $\mathbf{F}_{friction}$ is to the left. The second thing that happens is probably not so familiar but is easy to observe. After the bounce, the ball spins CCW less rapidly, because $\mathbf{F}_{friction}$ counters the spin of the ball. Because of the properties of the basketball, the effect of $\mathbf{F}_{friction}$ is not large enough to reverse the direction of the spin. It is just able to slow the spin. If a lacrosse ball spinning CCW is dropped onto the floor, however, its spin reverses, becoming CW. This is because the lacrosse ball is significantly smaller and has less mass than a basketball and because the lacrosse ball is solid rather than hollow.

The deflection of a bouncing basketball caused by spin and the change of the spin are part of what expert dribblers use to cause deception. It is the other part of what dribblers do (more properly what the referees allow them to do) today that gives me heart burn. What I'm referring to is the contact of the basketball with the hand. In the old days, it was necessary for the hand to only contact the upper half of the ball during a dribble. In the language of physics, the normal force of the hand on the ball was not allowed to have an upward component; that is, $\mathbf{N}_{\text{hand on ball}}$ could only have a downward component. The fact that players are allowed to actually carry or "palm the ball" while dribbling significantly decreases my enjoyment of today's game.

Now that I've shown my age, let's get back to physics and try to analyze the bounce pass. There are many ways of throwing a bounce pass, two-hand, overhand, underhand, sidearm, and so on. If only a horizontal axis of rotation is considered, there are only three possibilities. The ball either has topspin, backspin, or no spin. The paths of typical topspin (open circles) and backspin (closed circles / dark spots) bounce passes that might be used in a game are shown in figure 5.12.

Each type of pass has its use. Clearly, according to figure 5.12, the topspin on the ball gives the basketball extra horizontal distance. There are occasions when this is useful. In general, though, the bounce pass with backspin is better, mainly because it is easier to catch. There are several reasons. First, the speed after the bounce is less than for the pass with a topspin. For the examples shown, the linear speed of the basketballs for both passes before the bounce is about the same, though the speed with which the backspin pass is thrown, 8 m / s (18 mph), is slightly smaller than the speed of the topspin pass, 8.5 m / s. After the bounce, the minimum linear speed (at the top of the bounce) of the pass with backspin is about 3 m / s while that for the ball with topspin is about 5 m / s. One consequence is that a bounce pass thrown with backspin leaves the floor in a more vertical direction. This is apparent from figure 5.12. This makes the backspin pass easier to catch since

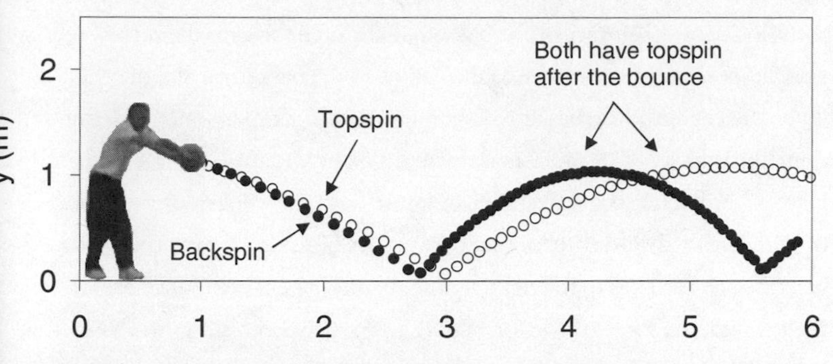

Figure 5.12. Bounce passes with topspin and backspin. The position of the basketball is shown every 1/60 second.

it bounces up to the receiver. Another reason that the ball with backspin is easier to handle is that it spins more slowly after the bounce. Both basketballs were thrown with roughly the same rotational speed, about 4 rotations per second for the ball with backspin and 5 rotations per second with topspin. The ball with an initial backspin has topspin after the bounce with a rotational speed of 4.6 revolutions per second. The ball with an initial topspin has topspin after the bounce with a rotational speed of 7.5 revolutions per second. The faster spinning basketball is more difficult to catch.

Even though a basketball with backspin is easier for the receiver to handle and in most cases easier to throw, I used a topspin bounce pass from time to time in a game. For example, if the person on defense was close and the receiver was farther away, it was sometimes useful to bounce the ball with topspin near the person on defense. The ball is more difficult to catch but sometimes it's the best or only way to get it there. My guess is that the bounce pass shown in figure 5.13 had backspin.

The physics of the bounce pass is interesting. I only casually mentioned that even though one basketball was thrown with topspin and the other was thrown with backspin both have topspin after the bounce, both basketballs are slowed by the bounce. Since this experiment was done very early in the preparation of the book, I was sure that the basketball with topspin would speed up after the bounce but it didn't. The reason that the basketball behaves as it does follows from the discussion in chapter 4. There we saw that the velocity of the surface of the basketball through the air is the sum of the velocity of the surface relative to the center of the basketball and the velocity of the center of the basketball. For a basketball only spinning CW(topspin), the velocity of the bottom is to the left relative to the center of the basketball. For the bounce pass shown in figure 5.12, the velocity of the center of the basketball dominates so that the velocity of the bottom of the basketball through the air is to the right. Consequently, when the bounce pass hits the ground, the force of friction is to the left. That both slows the basketball and increases the rate of topspin. It is possible to put so much topspin on the ball that the surface speed of the ball is greater than the linear speed of the ball, so that the bounce will cause the ball to speed up and slow the spin. However, this situation does not usually occur during a game.

Looking back through the rest of this book, I see one topic that appears in every chapter—the spin of the basketball. It's interesting to speculate that whoever controls the spin controls the game.

Figure 5.13. Steve Nash #13 of the Dallas Mavericks bounce-passes the ball to Michael Finley past Byron Russell #3, John Stockton #12, and Karl Malone #32 of the Utah Jazz circa 2001 at the American Airlines Arena in Dallas, Texas. © 2006 Brian Spurlock / NBAE / Getty Images

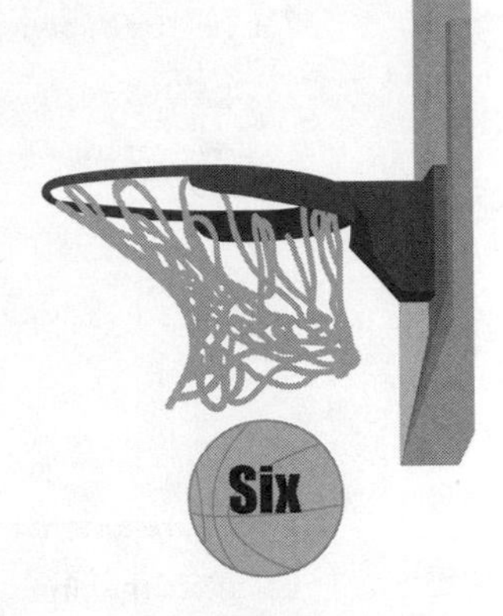

Hang Time

Gus Johnson, who was a star with the Baltimore Bullets, introduced me to the concept of hang time. We were playing in an exhibition game. Okay, he was playing and I was doing whatever it was that I did. In this case, I was moving down the left side of the key. He had the ball at about the foul line, then leaped toward the basket for a dunk. He seemed to float in air then slammed the ball through the hoop. I still have a mental picture of him hanging or floating in air, apparently defying gravity and the laws of physics. The illusion of hang time has the following explanation.

Suppose that a player jumps and that only gravity acts when the player is in the air. Physics says that the amount of time that the player is in the air is determined only by what happens in the vertical direction. That is because gravity acts downward. When the player jumps, she is initially moving upward at a relatively high speed. That implies that in the first part of the jump the player moves a relatively large vertical distance in a short amount of time. Consequently, it takes a relatively short amount of time to reach half the maximum height of the jump. Because of gravity, the player's vertical speed decreases while she is moving upward. This implies that toward the top of the jump she is moving upward more slowly than at the beginning of the jump. At the top of the jump, the vertical speed of the player is zero. Consequently, it takes a relatively long time to go from half the maximum height to the maximum height. The longer time spent in the upper half of the jump explains the illusion of hang time.

We can be more specific. The following kinematics equations,

$$v_y^2 - v_{0y}^2 = 2a_y y \quad (6.1)$$

and

$$v_y - v_{0y} = a_y t, \quad (6.2)$$

which are taught in essentially all general physics courses, can be used to calculate the times associated with the vertical part of a jump. We assume that y is height above the ground, v_y is the velocity upward (in the y direction) at any time, t, and that $a_y = -9.8\ \text{m/s}^2$, the acceleration of gravity. The minus sign on a_y indicates that the acceleration is downward. v_{oy} is the initial ($t = 0$) upward velocity that occurs at $y_o = 0$.

Suppose that a player jumps to a height b ($y_{top} = b$). Since we know that the vertical speed of the player is zero at the top of the jump, ($v_{y,top} = 0$), equation (6.1) tells us that $v_{0y} = \sqrt{19.6b}$. We can plug all of this into equation (6.2) with the result that

$$t_{top} = \frac{\sqrt{19.6b}}{9.8} = \frac{4.43\sqrt{b}}{9.8}$$

Next, if $y_{mid} = b/2$, equation (6.1) tells us that $v_{y,\,mid} = \sqrt{9.8b}$ so that

$$t_{mid} = \frac{1.3\sqrt{b}}{9.8}.$$

From this we calculate that $t_{mid} = 0.29t_{top}$. This tells us that a jumping player spends only 29% of the time in the bottom half of the jump. Correspondingly, a player must spend 71% of the time in the upper half of the jump. These are the numbers that show that the apparent illusion of hanging or floating in air is primarily due to the increased amount of time that a player spends in the top half of the jump. This is summarized in figure 6.1.

The illusion is enhanced by the horizontal motion of the player. The fact that 71% of the time is spent in the upper half of the jump implies that 71% of the horizontal distance traveled occurs while the player is in the upper half of the jump. As a consequence, the player travels most of the horizontal distance while in the upper half of the jump.

During the jump associated with a stuff or slam dunk, there can be another contribution to the illusion. In this case the basketball is usually raised then lowered through the basket. Consequently, the shape of the body/basketball system often changes a small amount. The center of mass of the body/basketball system must continue on its usual approximately parabolic trajectory, but the trajectory of the body itself can be different. The force of the basketball on the hands can change the trajectory of the body. Conversely, the force of the hands on the basketball can change the trajectory of the basketball. Raising the ball (and arm) results in a lowering of the body. When the ball is lowered, the body rises. If the raising of the ball occurs on the way up and the lowering occurs on the way down, the result is a leveling off of the trajectory of the body. This enhances the illusion of floating.

Figure 6.1. Josh Smith of the Atlanta Hawks takes off in the Sprite Rising Stars Slam Dunk competition on February 19, 2005. ©2006 Ezra Shaw / NBAE / Getty Images

Next, we'll consider the jump itself. The jump is also the first and, to some of us, the most critical phase of shooting a jump shot. It goes without saying that the jump is also important in rebounding. In keeping with my obsession with shooting, we'll start with the jump associated with the jump shot. In our analysis of the jump shot, we have been working backward. We started by considering the flight of the ball. Second, we analyzed the release of the ball. Third, we developed a model for the launch of the ball by means of the hand-basketball contact. Fourth, we considered the proper configuration of the rest of the body during a shot. Finally, we will analyze the interaction of the floor with the shooter.

I used the Vernier Force Plate® again. In this case, measurements of the force versus time were made as I jumped off the force plate while shooting a jump shoot. The force that was measured is the force that I exerted on the force plate. Again, by N3L, that force is equal and opposite to the force of the force plate on me. The results are shown by the solid line in figure 6.2. Simultaneously, a video was made of the jump shot. Subsequently, the Videopoint 2.5® software was used to analyze the video. The video and the data from the force plate were synchronized with the time when the force plate was first contacted. The time when I first contacted the force plate was 2.4 s and that was used as the reference. The vertical velocity of my head is shown by the solid circles in figure 6.2. The velocity data are similar to the data shown in figure 3.4.

I stepped onto the force plate at 2.4 s. From 2.4 s to about 4.0 s, I was settling into a standing position on the force plate. I remained still from about 4.0 s to about 4.9 s. The purpose of standing still was to get a good reading equal to my weight. This provided a check of the calibration of the force plate. It also made it easier to include gravity in the calculation of impulse. That calculation will be described shortly.

At about 4.9 s, the jump begins. The initial phase of the jump is characterized by a dip/decreasing force labeled knee bend in the figure. The knee bend takes place from about 4.9 s to about 5.1 s. During this phase of the jump my center of mass is lowering (accelerating downward) because my knees are bending. Since I am accelerating downward, the

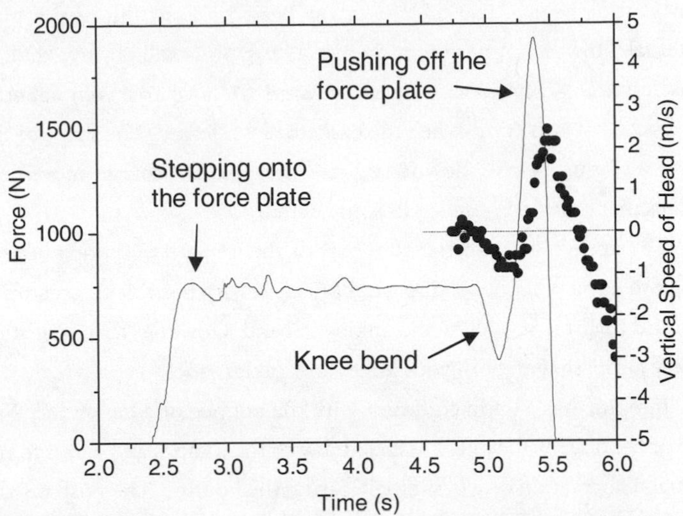

Figure 6.2. Plot showing the force versus time of the floor on and vertical speed versus time of the author during a jump shot from around the foul line

upward force of the force plate on me is less than the downward force of gravity. That makes the reading on the force plate less than my weight. The purpose of the knee bend is to reconfigure the body so that it can exert a large force on the force plate. That enables the force plate to exert a large force on me, thereby causing the jump. The details of the force and the jump are as follows. Beginning at 5.1 s the knees begin to straighten, and so on. Correspondingly, the force begins to increase rapidly. From 5.1 s to 5.2 s I am still accelerating downward because it takes a while to build up to a force greater than gravity. The downward acceleration over the range 4.9 to 5.2 s is apparent from the velocity of my head. The velocity becomes increasingly negative over this time interval. For times greater than 5.2 s the upward force of the force plate is greater than the downward force of gravity and continues to increase rapidly until it reaches a maximum at 5.35 s. It then decreases rapidly from 5.35 s to about 5.5 s when I lose contact with the force plate. However, I only accelerate upward from 5.2 s to 5.45 s. During this time interval the velocity of my head increases upward. From 5.45 s to 5.5 s the force is less than gravity so the downward acceleration begins. The velocity of my head decreases during this time interval. However, my velocity from 5.45 to 5.5 s is still positive though it is decreasing. My velocity is positive until 5.75 s when it is momentarily zero. That is the time when the basketball is released. The shape of the force versus time curve shown in figure 6.2 is typical of countermovement jumps that have been studied in some detail.[1]

The knee bend is an essential part of the jump. If you don't believe it, try to jump without bending your knees. Yes, it's possible to get a small upward thrust if you allow the ankles to bend. However, that's cheating so try jumping with both your knees and ankles locked.

Jumping has a lot in common with the bounce of a basketball. The dip or bending of the knees associated with the jump is analogous to the initial deformation of a basketball during the bounce. The push off the force plate during a jump is similar to the restoration of the shape of the basketball during the second half of the bounce. Both forces exhibit a maximum. The analogy is not perfect. In the case of the bounce, the speed of the basketball is zero when the force is a maximum. The speed is not zero when the force is a maximum for the jump. The real reason for developing the analogy is to show what it would take to reveal why some people can jump higher than others. To learn why some basketballs bounce higher, it was necessary to dissect the basketball. That revealed the role of the materials-dependent elastic force in the bounce. In order to learn why some people jump higher, it would be necessary to dissect the body. One possibility is to cut the player off at the knees, figuratively speaking. That would reveal the role of the forces caused by the muscles, tendons, and ligaments in the jump. We'll leave that to the biomechanics people and just concentrate on the basic physics.

As for a bouncing basketball, the impulse defined by equation (5.1) is of interest. For the jump represented in figure 6.2 the upward impulse from about 4 s to 5.5 s was found to be 1,280 N s. The downward impulse due to gravity over this time interval is 1,150 N s. This gives a net upward impulse of 130 N s. According to equation (5.1), my center of mass should have been traveling upward at about 1.7 m/s when I lost contact with the force plate. The data shown in figure 6.2 for the upward speed of my head are in agreement with this prediction.

Needless to say, my ability to jump is pathetic. In my playing days I could touch the rim with the palm of my hand but I never could quite dunk. It's now 40 years later, however, and I've had an operation on each knee so it's impossible for me to demonstrate the jumping ability of today's (or even yesterday's) player. Fortunately, a talented Navy basketball player, Calvin White, agreed to help. We made videos of his jumps and measured the force versus time as he jumped off the force plate on his way to a two-hand stuff. The results of the force measurements are shown by the solid line in figure 6.3. The scale for the force is on the left. A picture of the two-hand stuff is also shown in the figure. The arrow below the picture points to the time that the picture was taken. The position of Calvin's head every 1/60 s was digitized by using the Videopoint 2.5® software. The vertical velocity was calculated from the data and is shown in figure 6.3 by the circles. The axis for the vertical velocity is on the right. Again, the data from the force plate and the video were synchronized using the time when the force plate was first contacted.

The force of the floor on Calvin began at about 3.1 s and rose to a maximum at about 3.5 s. The rise was not smooth. The small features on the curve between 3.1 s and 3.35 s reflect various aspects of Calvin's jump. Between 3.1 s and 3.35 s Calvin was transitioning to the force plate. He first contacted the force plate with the left foot then brought the right foot onto the force plate. At 3.35 s, he began bending his knees. What also happens at 3.35 s is that he began moving his head upward. This is indicated by the positive values of the velocity beginning at about 3.35 s. The data show that the velocity reaches a maximum at a later time than the force. As seen by a close inspection of figure 6.2, the same behavior was observed for my jump shot. Even though the force is decreasing at times greater than about 3.5 s, it is still positive. As long as the upward force of the force plate on Calvin is greater than the downward force of Calvin's weight, Calvin accelerates upward.

The force decreased rapidly from 3.5 s to about 3.6 s when Calvin left the force plate. Calvin continued upward until about 3.9 s when he began

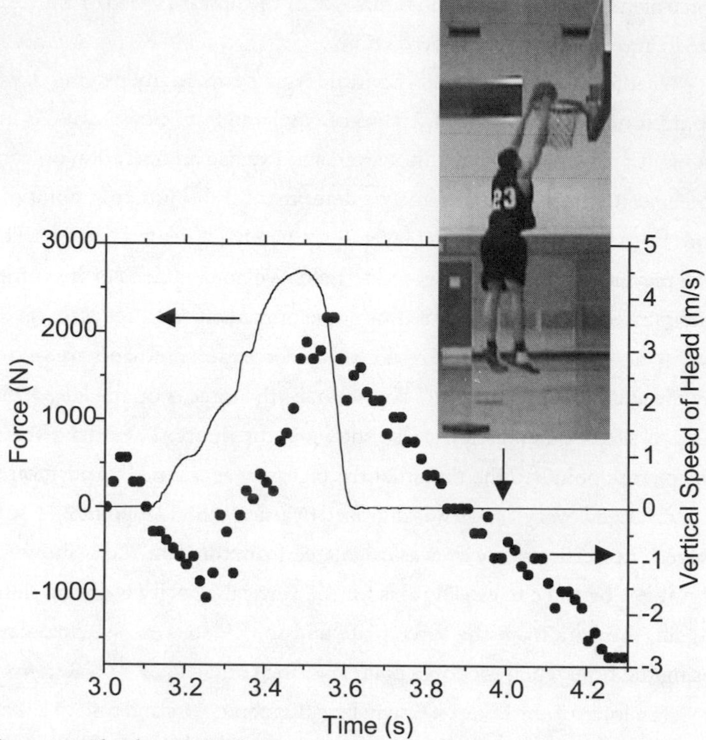

Figure 6.3. Plot showing the force versus time of the floor on and vertical speed versus time of Calvin White associated with a two-hand stuff

to move downward. The arrow below the picture superimposed on the graph indicates the time of the picture was about 4.0 s. I chose that frame from the video because it seemed to be the closest to the beginning of the two-hand stuff. I found it interesting that the vertical velocity is negative at that time. That means that the stuff is initiated when Calvin is moving downward. I closely watched a slam dunk contest on TV the other day and noticed that most dunks are initiated when the player is on the way down. That's different from jump shots where the basketball is released at the top of the jump.

Needless to say, there's quite a bit of difference between Calvin's and my ability to jump. He was able to exert a maximum force of about 2,800

N (630 lbs) on the floor while the best that I could do was about 1,800 N (400 lbs). A small amount of the difference is because he weighs about 1.3 times as much as I do. However, the key factor is that he generated a maximum force of about 2.8 times his weight while the best that I can do is only about 2.3 times my weight. In fact, Calvin can probably do even better than that since his goal was to achieve a two-hand stuff and not to generate the most force possible.

Our discussion of the jump has been relatively academic. It suggests an easy method for testing players, though. It would be straightforward to set up a force plate and video in the gym. They could be used to evaluate a player's ability to jump. A measurement of the impulse that a player is able to deliver to the force plate/floor is a reliable indication of the player's ability to jump. The simultaneous video would allow evaluation of a player's jumping technique. This could be done in the off-season for training or during the season for extra evaluation and training. It also might be useful at tryouts to help decide whom to cut.

I'll end the book with a discussion of what might happen if Calvin grabbed the rim on the way down from dunking the ball. A spectacular example of what might happen is shown in figure 6.4.

I hate to burst the bubble of those who shatter backboards but it isn't a measure of the strength of the player. Backboards don't usually shatter because large forces are applied. They shatter because of how a force is applied, and that force can be small. For example, the first smashing of the glass in professional basketball occurred at the Boston Arena on November 5, 1946 when a basketball hit the rim during warmups.[2] It was a Basketball Association of America game between the Boston Celtics and the Chicago Stags. The NBA was known as the Basketball Association of America until 1949. The person who shot the ball was Chuck Connors, who was playing for the Celtics and who later became star of *The Rifleman* TV series. The shot was a two-hand set shot from 15 to 20 feet. What happened was that a worker had not installed a piece of protective rubber between the basket and the backboard. This allowed the basket to exert a force on the backboard along an edge. The rear windows of some

Figure 6.4. Robert "Tractor" Traylor's smashing debut for the University of Michigan on November 26, 1996. The Wolverines defeated Ball State University 87 to 63. Jack Gruber, *Detroit News*

cars are made from the same material as backboards. I was waiting in line at the local emissions testing facility and happened to be looking at another car. I saw the rear window of the car "spontaneously" disintegrate in the same spectacular manner as a backboard. The lady who owned the car said that the window had just been replaced as part of a major repair that included bodywork. I suggested to her that the window had probably been installed incorrectly. My guess is that the problem was with the rubber seal. It probably allowed the frame of the car to exert a force on the window along the edge.

Both the backboard and the rear window were made from tempered or toughened glass. This kind of glass is more resistant to impact than ordinary glass. It usually shatters when something strikes or pushes on the glass at an edge, though it can also fail if the glass is bent enough. Tempered glass is resistant to impact and sensitive at the edge because of the way that it is made. One way to make tempered glass is to heat it about 150°C (270°F) above a special temperature, known as the glass transition temperature.[3] For many types of glass, that requires heating to about 650°C (1,200°F). The glass is then quickly removed from the heat and the large surfaces are rapidly cooled, often by air blasts. Consequently, the surfaces cool rapidly and the interior cools more slowly. Glass that is formed by rapid cooling occupies more volume (is less dense) than glass that is formed by cooling slowly. Consequently, the interior pulls inward on the outer surface and vice versa. This puts the outer surface in compression and the inner surface in tension. There is a delicate balance of the internal forces in tempered glass. A backboard shatters when the balance is upset. Alternatively, it can be said that there is lots of stored energy in tempered glass. From this point of view, a backboard shatters when the energy stored in the compressed surface and tensioned interior is released.

In general, glass can only break when it is under tension, for example, the edges of a microcrack are pulled apart. The compression in the large surface of tempered glass makes it much more difficult to break because the compression must be overcome before tension can be applied. The result is a glass that is four or five times as strong as ordinary glass. This ensures that a basketball traveling with ordinary speeds cannot break the backboard by colliding with it. However, the edges of the glass are different. The interior of the glass, which is already under tension, is accessible via the edges. When the edge/interior is perturbed in such a way as to increase the size of a microcrack or introduce a crack, the delicate balance of forces is upset and a crack begins to propagate. Since the crack, itself, further upsets the balance, it generates other cracks. Apparently, cracks can accelerate to speeds up to about one third the speed of sound in the material.[4] For typical glass this is about 1,500 m/s (3,400 mph). At this speed, it takes only about 0.001 s for a crack to travel the length of a backboard. Because the cracks propagate at such high speeds, the backboard appears to shatter instantaneously.

That's about it, at least for now. There is a lot more that I'd like to write about but the book stops here. I hear the fat lady singing.

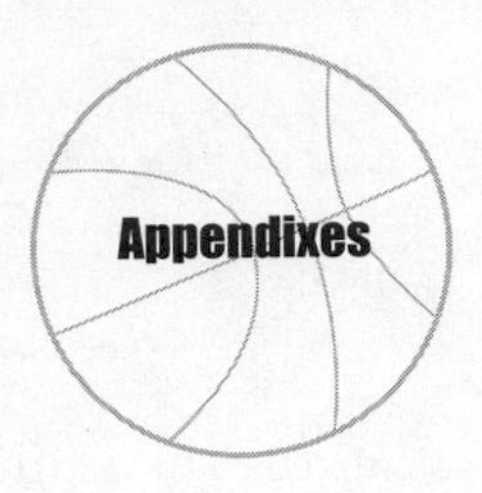

I. The Drag Force

Depending on the speed, size, and shape of an object, the drag force has different descriptions. What works well for the magnitude of the drag force on a basketball traveling with speeds less than about 10 m/s is equation (1.5). Using a density of air, ρ_{air}, of 1.22 kg/m^3 and a drag coefficient, C_{drag}, of 0.25 we obtain the following working equation for the magnitude of the drag force on a men's basketball

$$F_{drag} = 0.0136\,\upsilon^2\,. \qquad \text{(I.1)}$$

For a women's basketball, we get

$$F_{drag} = 0.0127\upsilon^2\,. \qquad \text{(I.2)}$$

υ has units of meters per second and F_{drag} has units of newtons.

II. The Magnus Force

The Magnus force is given by equation (1.6). The Magnus coefficient, C_{Magnus}, was set equal to 1.08 per revolution. The diameter of a basketball, D, is 0.24 m for a men's basketball and 0.23 m for a women's basketball. The value of the rotational speed, ω, is usually set equal to what was measured. This leads to the following working equation for a men's basketball

$$F_{Magnus} = 0.0180\omega\upsilon. \qquad \text{(II.1)}$$

For a women's basketball, we get

$$F_{\text{Magnus}} = 0.0162\omega v. \qquad \text{(II.2)}$$

III. Trajectory Calculations

First, the velocities were calculated as time increases. In the horizontal direction (x direction), only the drag and Magnus forces act so Newton's Second Law gives

$$m\frac{dv_x}{dt} = -C_{\text{drag}}\rho A v v_x - C_{\text{Magnus}}\rho D^3 \omega v_y. \qquad \text{(III.1)}$$

All equations are written with ω positive for a basketball with backspin. For a basketball with topspin ω is negative. This leads to the working equation for a men's basketball

$$\Delta v_x = (-0.0223\, v v_x - 0.0295\,\omega v_y)\Delta t \qquad \text{(III.2)}$$

and

$$\Delta v_x = (-0.0223\, v v_x - 0.0285\,\omega v_y)\Delta t \qquad \text{(III.3)}$$

for a women's basketball.

In the vertical direction (y direction), all four forces act so Newton's Second Law gives

$$m\frac{dv_y}{dt} = F_{\text{gravity}} + F_{\text{buoyant}} - C_{\text{drag}}\rho A v v_y + C_{\text{Magnus}}\rho D^3 \omega v_x. \qquad \text{(III.4)}$$

Again, the sign for the Magnus force is reversed for a basketball with topspin. This leads to the working equation for a men's basketball

$$\Delta v_y = (-9.66 - 0.0223\, v v_y + 0.0295\,\omega v_x)\Delta t \qquad \text{(III.5)}$$

and

$$\Delta v_y = (-9.66 - 0.0223\, v v_y + 0.0285\,\omega v_x)\Delta t \qquad \text{(III.6)}$$

for a women's basketball.

These equations were used to predict the velocity as follows. First, initial ($t = 0$) values of v_x, the horizontal component of the launch speed, and v_y, the vertical component of the launch speed, were chosen then Δv_x and Δv_y were calculated for a small time interval, typically 0.001 s. The changes were added to the initial values to find the values of v_x and v_y at 0.001 s. The values of Δv_x and Δv_y were then calculated using the values of v_x and v_y at 0.001 s. The new values of Δv_x and Δv_y were then added to the values of v_x and v_y at 0.001 s to find the values of v_x and v_y at 0.002 s. This process was repeated using an Excel spreadsheet until the values of v_x and v_y at the final time were calculated.

Once the velocities were calculated, the horizontal and vertical positions, x and y, were determined as follows. The definitions of the average velocities

$$\Delta x = \overline{v_x} \Delta t \qquad \text{(III.7)}$$

and

$$\Delta y = \overline{v_y} \Delta t \qquad \text{(III.8)}$$

were used to calculate the changes in the horizontal and vertical positions, Δx and Δy, for the same time interval used for the velocity calculations. Next, an initial ($t = 0$) position was chosen. The changes were added to the initial values for each time interval in a manner similar to that used for the velocities and so on and the result was that both x and y were calculated as time increases.

IV. Calculation Setup

The setup for the numerical calculations with a women's basketball is shown in figure A.1.

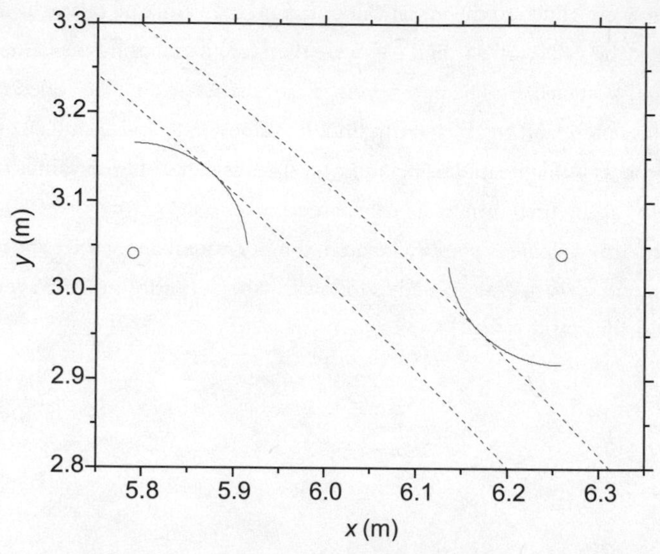

Figure A.1. Setup for numerical calculations for a women's basketball

The small circles are the rim and the quarter circles are drawn at the radius of the rim plus the radius of the basketball. The dashed lines are the extreme trajectories of the center of the basketball for a launch angle of 50.8°. The upper dashed line has a launch speed of 6.80 m/s and the lower dashed line has a launch speed of 6.71 m/s. When a dashed line touches a quarter circle, as shown for the two trajectories, the basketball touches the rim.

V. Bounce Angles

To give a more detailed treatment of the bounce of a basketball from a rim, it is useful to describe the velocity via its angle below the horizontal, θ_{VEL}, when it contacts the rim. That angle is shown in Figure A.2.

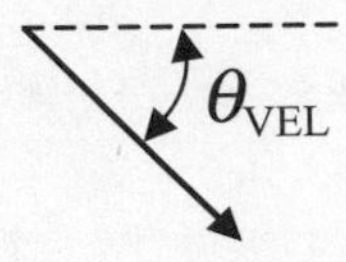

Figure A.2. Sketch showing the angle of the velocity below the horizontal when the ball contacts the rim, θ_{VEL}.

We begin by describing the bounces where θ_{PC} is between 0° and 45° as shown in figure 4.2b. What we find is that a basketball that hits the rim with θ_{VEL} between 0° and $2\theta_{PC}$ bounces upward and to the left. If the ball approaches the rim with θ_{VEL} between $2\theta_{PC}$ and 90°, it bounces downward and to the left. When $\theta_{VEL} = 2\theta_{PC}$, the ball leaves the rim horizontally to the left.

Next, we describe the bounces where θ_{PC} is between 45° and 90° as shown in figure 4.3b. What we find is that a basketball that hits the rim with θ_{VEL} between 0° and $2\theta_{PC}-90°$ bounces upward and to the right. If the ball approaches the rim with θ_{VEL} between $2\theta_{PC}-90°$ and 90°, it bounces upward and to the left. When $\theta_{VEL} = 2\theta_{PC}-90°$, the ball leaves the rim vertically upward.

Next, we describe the bounces where θ_{PC} is between 90° and 135° as shown in figure 4.4b. What we find is that a basketball that hits the rim with θ_{VEL} between $\theta_{PC}-90°$ and $2(\theta_{PC}-90°)$ bounces upward and to the right. If the ball approaches the rim with θ_{VEL} between $2(\theta_{PC}-90°)$ and 90°, it bounces downward and to the right. When $\theta_{VEL} = 2(\theta_{PC}-90°)$, the ball leaves the rim horizontally to the right.

Finally, for bounces where θ_{PC} is between 135° and 180° as shown in figure 4.4c, all bounces are downward and to the right. The smallest possible value of θ_{VEL} is $\theta_{PC}-90°$, and the largest is 90°.

VI. Coefficient of Restitution

In the collision of a basketball with the floor the coefficient of restitution, *e*, is defined as

$$e = -\frac{v_{\text{floor, after}} - v_{\text{ball, after}}}{v_{\text{floor, before}} - v_{\text{ball, before}}}. \qquad \text{(VI.1)}$$

The velocity of the floor before the collision, $v_{\text{floor,before}}$, is assumed to be zero. Because of conservation of momentum, the velocity of the floor after the collision, $v_{\text{floor,after}}$, cannot be exactly zero, but if the floor is connected to the earth it is very, very close to zero. Consequently, we rewrite the previous equation as

$$e = -\frac{v_{\text{ball, after}}}{v_{\text{ball, before}}}. \qquad \text{(VI.2)}$$

The reason for the minus is to make e positive. This is necessary if, for example, we let the velocity of the ball after the collision, $v_{\text{ball,after}}$, be positive since it is moving upward and the velocity of the ball before the collision $v_{\text{ball,before}}$, be negative since it is moving downward. A ball that doesn't bounce at all has a coefficient of restitution of 0 since the speed of the basketball after the collision is 0. A perfect basketball has a coefficient of restitution of 1 because it will have the same speed before and after the collision.

Until the past few years, it has been difficult to measure the speed of the ball before and after collision. Consequently, for practical purposes, equation (VI.2) is usually transformed using the standard general physics equation relating the speed acquired by a ball falling from rest a distance h

$$v = \sqrt{2gh}\,. \qquad \text{(VI.3)}$$

This equation also holds for any constant acceleration such as would be caused by both gravity and the buoyant force. Since this equation also can be used to calculate the maximum height, h, that a ball rises if it is thrown upward with a speed, v, equation (VI.2) is often rewritten as

$$e = \sqrt{\frac{h_{\text{ball, after}}}{h_{\text{ball, before}}}}. \qquad \text{(VI.4)}$$

For example, if the ball doesn't bounce at all, the height is zero, and thus the ball again has a coefficient of restitution of 0. The coefficient of restitution calculated for other bounces using the height equation is wrong, however, and we know the reason why—air resistance. Air resistance slows down a basketball both on the way down (before the bounce) and on the way back up. I calculated the effect of air resistance and found that a perfect basketball dropped from a height of 1.8 m behaves as though it has a coefficient of restitution of about 0.96 rather than the value of 1. Consequently, although the height of the bounce is a valid indicator of the level of inflation of a basketball, it should not be used to calculate the coefficient of restitution by the standard equation involving the heights.

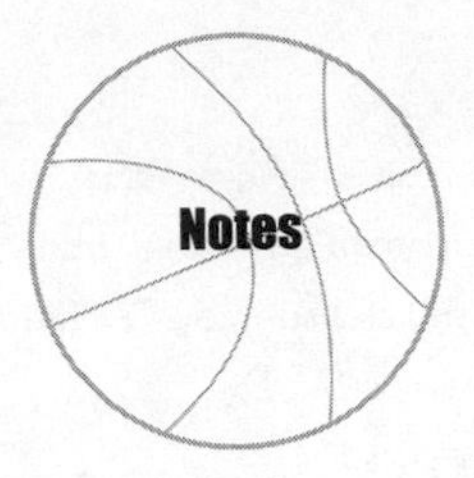

Chapter 1. The Final Four

1. I am assuming that the basketball team at the California Institute of Technology will take a few minutes and read the book. In his column in the January 9, 2006 issue of *Sports Illustrated* (p. 74), Rick Reilly points out that, as of that date, the Caltech Beavers had lost its 183rd straight NCAA game. To paraphrase Reilly: Go Tech, Cure Cancer.
2. The quote was used in this form and made famous by San Antonio sports broadcaster Dan Cook during a television newscast in April 1978. The newscast occurred after the first game between the San Antonio Spurs and the Washington Bullets during the 1977–1978 NBA playoffs. The quote appears to have existed in related forms for many years prior to 1978.
3. Another argument showing why this is not an action–reaction pair is that if the floor is accelerating, as would be the case if the floor were the floor of an elevator just starting or stopping, the force of gravity and the force of the floor on the ball would no longer be equal and opposite. An action–reaction pair must always be equal and opposite.
4. *Philosophiae Naturalis Principia Mathematica*, I. Newton, lib. 2, prop. 33 (1687). (Citation from Barkla and Auchteronie as quoted in note 5.)
5. According to Barkla and Auchterlonie ("The Magnus or Robins Effect on Rotating Spheres," *Journal of Fluid Mechanics*, 42 (1971): 437–447), in 1742 B. Robins demonstrated the existence of a transverse aerodynamic force on a rotating sphere (*New Principles of Gunnery*, ed. Hutton (1805), 210, but first printed in 1742). That is more than 100 years before the publication of the paper by Magnus in 1853 ("Ueber die Abweichung der Geschosse, und eine auffallende Erscheinung bei rotirenden Körpern," *Poggendorfs Annalen der Physik and Chemie* 88 [1853]: 1) on the existence of such forces on rotating cylinders. Nonetheless, the transverse force on any moving, spinning object is now known as the Magnus effect.

6. Robert K. Adair, *The Physics of Baseball*, 3rd ed. (New York: HarperCollins, 2002).
7. Adair, *The Physics of Baseball.*
8. G. Ireson, "Beckham as Physicist?" *Physics Education* 36 (2001): 10–13. In his paper, Ireson cites the 1999 web site of J. P. Carini (http://carini.physics.indiana.edu/E105/spinning–balls.html) as a source of equation (1.6).
9. Ireson, "Beckham as Physicist?"
10. Ireson, "Beckham as Physicist?"

Chapter 2. Projectile Notion

1. Gary M. Pomerantz, *Wilt, 1962* (New York: Crown, 2005).
2. J. G. Hay, *The Biomechanics of Sports Techniques*, 4th ed. (Englewood Cliffs, NJ: Prentice Hall, 1993).
3. Hay, *Biomechanics of Sports Techniques*. P. J. Brancazio, "Physics of Basketball," *American Journal of Physics*, 49 (1981): 356–365.
4. The key is the area of the floor directly in front of the basket. I remember when the lines and curves marked on the floor in the vicinity of each hoop actually resembled a key hole. The rectangle and semicircle that are painted on most courts today are much less interesting. The top of the key is the point on the semicircle furthest from the basket.
5. Brancazio, "Physics of Basketball."
6. Brancazio, "Physics of Basketball."
7. Three–point shots (shots from beyond the three–point line) did not exist while I was playing. According to Joe Onderko, the statistics for my sophomore year are incomplete. I was able to piece together the foul–shooting statistics using the data supplied by Onderko and the information for my sophmore year in *The Glory Years* by Dick Minteer. The field goal percentage is for freshman, junior, and senior years only.
8. There is an error in the Individual Collegiate Records section of the *2006 NCAA® Men's Basketball Records Book* (Indianapolis, IN: The National Collegiate Athletic Association, 2005). It omits the record by Enfield. That has been corrected in the *2007 NCAA® Men's Basketball Records Book*, (Indianapolis, IN, The National Collegiate Athletic Association, 2006).
9. *2006 NCAA® Women's Basketball Records Book* (Indianapolis, IN: The National Collegiate Athletic Association, 2005).

Chapter 3. Nothing But Net

1. My memory of both games is helped considerably by the video on the DVD that comes packaged with the book, *Not Till the Fat Lady Sings* by Les Krantz, (Chicago, IL: Triumph Books, 2003).
2. Worthy went on to an outstanding 12–year NBA career. He played ten of the years with the Los Angeles Lakers.
3. Krantz, *Not Till the Fat Lady Sings*.
4. P. J. Brancazio, "Physics of Basketball," *American Journal of Physics*, 49 (1981): 356–365.
5. *Sports Illustrated*, March 24, 1986.
6. We refer to an outside shot as a shot from a distance to the hoop of greater than about 4.2 m (13.8 ft). An outside shot is a shot from beyond the distance from the foul line to the hoop.
7. For the remainder of this book, a one–hand set shot is referred to simply as a set shot. As very old timers will remember, there are also two–hand set shots. Although two–hand set shots are useful from very great range, midcourt and beyond (the last shot that I made at Buzz Ridl Field House at Westminster College during a game was a two–hand set shot from midcourt), I see no advantage in using it for shots from ordinary distances. Consequently, two–hand set shots are not considered further in this book.
8. Robert K. Adair, *The Physics of Baseball*, 3rd ed. (New York: Harper Collins, 2002).

Chapter 4. Basket Case

1. In this book the rim is defined as the thin metal rod. The hoop is the result of bending the metal rod into a circle. The basket is the hoop plus net.
2. G. Wahl, "It All Starts Here," *Sports Illustrated*, February 17, 2003, 56. This article was pointed out to me by Frank Oslislo during one of our chats at the half–time of Navy games. We played a lot of basketball together when we were young. Coincidently, Frank and his college team, Millersville University, also played in the 1967 NAIA tournament. Frank's high school coach was Pete Carril.

Chapter 5. That's the Way the Ball Bounces

1. Note to the Internal Revenue Service and my wife: Purchasing a big–screen TV and watching lots of basketball games were necessary for writing this book.

2. Many of the features of this model were suggested by my wife, Dr. Mary Wintersgill. The development also benefited greatly from discussions with Dr. Charles A. Edmondson.
3. Physicists have a way of dealing with the height of bounce quantitatively. They define something called the coefficient of restitution to characterize bounces. A discussion of the coefficient of restitution in a sports context is given in the book *The Biomechanics of Sports Techniques*, 4th ed., by J. G. Hay (Englewood Cliffs, NJ: Prentice Hall, 1993). The mathematical details and some comments about the coefficient of restitution are given in appendix VI.
4. The occasion was a reception for recipients of national NAIA awards in Kansas City, Missouri. I won the Emil S. Liston Award that year.
5. R. Cross, "Grip–Slip Behavior of a Bouncing Ball," *American Journal of Physics* 70 (2002): 1093–1102.

Chapter 6. Hang Time

1. N. P. Linthorne, "Analysis of Standing Vertical Jumps Using a Force Platform," *American Journal of Physics* 69 (2001): 1198–1204; R. Cross, "Standing, Walking, Running and Jumping on a Force Plate," *American Journal of Physics* 67 (1999): 304–309.
2. K. Banks, *The Unofficial Guide to Basketball's Nastiest and Most Unusual Records (Unofficial Guide)*, (Berkeley, CA: Greystone Books, 2005).
3. H. G. Pfaender, *Schott Guide to Glass* (New York: Van Nostrand Reinhold, 1983). J. S. Amstock, *Handbook of Glass in Construction* (New York: McGraw–Hill, 1997).
4. C. J. Phillips, *Glass: The Miracle Worker* (New York: Pitman, 1941). W. S. Ellis, *Glass* (New York: Avon Books, 1998).

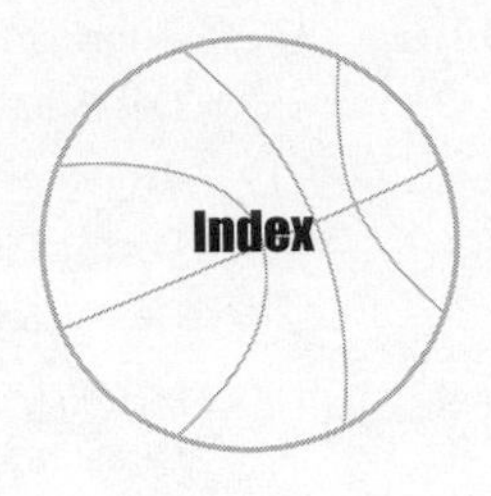
Index

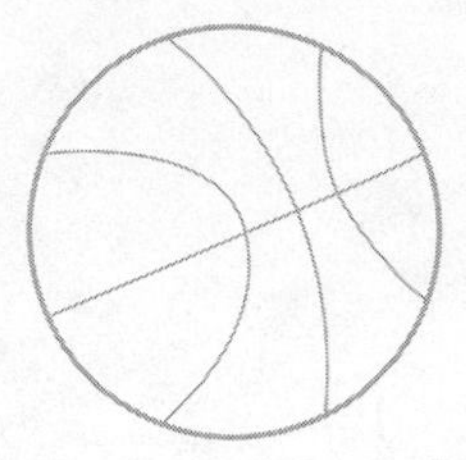

John J. Fontanella is a professor of physics at the United States Naval Academy. He currently has a research grant from the Office of Naval Research and has recently completed work funded by the National Science Foundation. He has published widely in the field of condensed matter physics. As of the writing of this book, he holds the single-game and single-season basketball scoring records at Westminster College (New Wilmington, Pa.). In 1967 he was an NAIA First Team All-American, College Sports Information Directors of America Academic All-American (College Division), and recipient of an NCAA Postgraduate Scholarship.